Kuang-Chao Fan (Ed.)

Design and Applications of
Coordinate Measuring Machines

MDPI

This book is a reprint of the Special Issue that appeared in the online, open access journal, *Applied Sciences* (ISSN 2076-3417) in 2016, available at:

http://www.mdpi.com/journal/applsci/special_issues/coordinate-measuring-machines-2016

Guest Editor
Kuang-Chao Fan
Department of Mechanical Engineering, National Taiwan University
Taiwan

Editorial Office
MDPI AG
St. Alban-Anlage 66
Basel, Switzerland

Publisher
Shu-Kun Lin

Senior Assistant Editor
Yurong Zhang

1. Edition 2016

MDPI • Basel • Beijing • Wuhan • Barcelona • Belgrade

ISBN 978-3-03842-276-1 (Hbk)
ISBN 978-3-03842-277-8 (PDF)

Table of Contents

List of Contributors...V

About the Guest Editor..IX

Preface to "Design and Applications of Coordinate Measuring Machines".............XI

Gaoliang Dai, Michael Neugebauer, Martin Stein, Sebastian Bütefisch and Ulrich Neuschaefer-Rube
Overview of 3D Micro- and Nanocoordinate Metrology at PTB
Reprinted from: *Appl. Sci.* **2016**, *6*(9), 257
http://www.mdpi.com/2076-3417/6/9/257..1

Rudolf Thalmann, Felix Meliand Alain Küng
State of the Art of Tactile Micro Coordinate Metrology
Reprinted from: *Appl. Sci.* **2016**, *6*(5), 150
http://www.mdpi.com/2076-3417/6/5/150..25

Hui-Ning Zhao, Lian-Dong Yu, Hua-Kun Jia, Wei-Shi Li and Jing-Qi Sun
A New Kinematic Model of Portable Articulated Coordinate Measuring Machine
Reprinted from: *Appl. Sci.* **2016**, *6*(7), 181
http://www.mdpi.com/2076-3417/6/7/181..43

Shih-Ming Wang, Yung-Si Chen, Chun-Yi Lee, Chin-Cheng Yeh and Chun-Chieh Wang
Methods of In-Process On-Machine Auto-Inspection of Dimensional Error and Auto-Compensation of Tool Wear for Precision Turning
Reprinted from: *Appl. Sci.* **2016**, *6*(4), 107
http://www.mdpi.com/2076-3417/6/4/107..59

Qiangxian Huang, Kui Wu, Chenchen Wang, Ruijun Li, Kuang-Chao Fan and Yetai Fei
Development of an Abbe Error Free Micro Coordinate Measuring Machine
Reprinted from: *Appl. Sci.* **2016**, *6*(4), 97
http://www.mdpi.com/2076-3417/6/4/97..80

Yin Tung Albert Sun, Kuo-Yu Tseng and Dong-Yea Sheu
Investigating Characteristics of the Static Tri-Switches Tactile Probing Structure
for Micro-Coordinate Measuring Machine (CMM)
Reprinted from: *Appl. Sci.* **2016**, *6*(7), 202
http://www.mdpi.com/2076-3417/6/7/202...97

**So Ito, Hirotaka Kikuchi, Yuanliu Chen, Yuki Shimizu, Wei Gao,
Kazuhiko Takahashi, Toshihiko Kanayama, Kunmei Arakawa and
Atsushi Hayashi**
A Micro-Coordinate Measurement Machine (CMM) for Large-Scale Dimensional
Measurement of Micro-Slits
Reprinted from: *Appl. Sci.* **2016**, *6*(5), 156
http://www.mdpi.com/2076-3417/6/5/156...112

Hiroshi Murakami, Akio Katsuki, Takao Sajima and Mitsuyoshi Fukuda
Reduction of Liquid Bridge Force for 3D Microstructure Measurements
Reprinted from: *Appl. Sci.* **2016**, *6*(5), 153
http://www.mdpi.com/2076-3417/6/5/153...139

**Rui-Jun Li, Meng Xiang, Ya-Xiong He, Kuang-Chao Fan, Zhen-Ying Cheng,
Qiang-Xian Huang and Bin Zhou**
Development of a High-Precision Touch-Trigger Probe Using a Single Sensor
Reprinted from: *Appl. Sci.* **2016**, *6*(3), 86
http://www.mdpi.com/2076-3417/6/3/86...153

Adam Gąska, Piotr Gąska and Maciej Gruza
Simulation Model for Correction and Modeling of Probe Head Errors in Five-Axis
Coordinate Systems
Reprinted from: *Appl. Sci.* **2016**, *6*(5), 144
http://www.mdpi.com/2076-3417/6/5/144...167

Jesús Caja, Piera Maresca and Emilio Gómez
A Model to Determinate the Influence of Probability Density Functions (PDFs) of
Input Quantities in Measurements
Reprinted from: *Appl. Sci.* **2016**, *6*(7), 190
http://www.mdpi.com/2076-3417/6/7/190...183

IV

List of Contributors

Kunmei Arakawa Engineering Department, MMC RYOTEC Corporation, Gifu 503-2301, Japan.

Sebastian Bütefisch Physikalisch-Technische Bundesanstalt, 38116 Braunschweig, Germany.

Jesús Caja Dpto. de Ingeniería Mecánica, Química y Diseño Industrial, ETS de Ingeniería y Diseño Industrial, Universidad Politécnica de Madrid, 28012 Madrid, Spain.

Yuanliu Chen Department of Finemechanics, Tohoku University, Sendai 980-8579, Japan.

Yung-Si Chen Department of Mechanical Engineering, Chung Yuan Christian University, Taoyuan 32023, Taiwan.

Zhen-Ying Cheng School of Instrument Science and Opto-electric Engineering, Hefei University of Technology, Hefei 230009, China.

Gaoliang Dai Physikalisch-Technische Bundesanstalt, 38116 Braunschweig, Germany.

Kuang-Chao Fan School of Instrument Science and Opto-electric Engineering, Hefei University of Technology, No.193, Tunxi Road, Hefei 230009, China; Department of Mechanical Engineering, National Taiwan University, l, Sec.4, Roosevelt Road, Taipei 10617, Taiwan.

Yetai Fei School of Instrument Science and Opto-electric Engineering, Hefei University of Technology, No.193, Tunxi Road, Hefei 230009, China.

Mitsuyoshi Fukuda Department of Mechanical Systems Engineering, Faculty of Environmental Engineering, The University of Kitakyushu, 1-1 Hibikino, Wakamatsu-ku, Kitakyushu, Fukuoka 808-0135, Japan.

Wei Gao Department of Finemechanics, Tohoku University, Sendai 980-8579, Japan.

Adam Gąska Laboratory of Coordinate Metrology, Cracow University of Technology, Kraków 31-155, Poland.

Emilio Gómez Dpto. de Ingeniería Mecánica, Química y Diseño Industrial, ETS de Ingeniería y Diseño Industrial, Universidad Politécnica de Madrid, 28012 Madrid, Spain.

Maciej Gruza Laboratory of Coordinate Metrology, Cracow University of Technology, Kraków 31-155, Poland.

Atsushi Hayashi Engineering Department, MMC RYOTEC Corporation, Gifu 503-2301, Japan.

Ya-Xiong He School of Instrument Science and Opto-electric Engineering, Hefei University of Technology, Hefei 230009, China.

Qiangxian Huang School of Instrument Science and Opto-electric Engineering, Hefei University of Technology, No.193, Tunxi Road, Hefei 230009, China.

So Ito Department of Finemechanics, Tohoku University, Sendai 980-8579, Japan.

Hua-Kun Jia School of Instrument Science and Opto-electric Engineering, Hefei University of Technology, Hefei 230009, China.

Toshihiko Kanayama Engineering Department, MMC RYOTEC Corporation, Gifu 503-2301, Japan.

Akio Katsuki Department of Mechanical Engineering, Faculty of Engineering, Kyushu University, 744, Motooka, Nishi-ku, Fukuoka 819-0395, Japan.

Hirotaka Kikuchi Department of Finemechanics, Tohoku University, Sendai 980-8579, Japan.

Alain Küng Federal Institute of Metrology METAS, 3003 Bern-Wabern, Switzerland.

Chun-Yi Lee Department of Mechanical Engineering, Chung Yuan Christian University, Taoyuan 32023, Taiwan.

Rui-Jun Li Anhui Electrical Engineering Professional Technique College, Hefei 230051, China; School of Instrument Science and Opto-electric Engineering, Hefei University of Technology, No. 193, Tunxi Road, Hefei 230009, China.

Wei-Shi Li School of Instrument Science and Opto-electric Engineering, Hefei University of Technology, Hefei 230009, China.

Piera Maresca Dpto. de Ingeniería Mecánica, Química y Diseño Industrial, ETS de Ingeniería y Diseño Industrial, Universidad Politécnica de Madrid, 28012 Madrid, Spain.

Felix Meli Federal Institute of Metrology METAS, 3003 Bern-Wabern, Switzerland.

Hiroshi Murakami Department of Mechanical Systems Engineering, Faculty of Environmental Engineering, The University of Kitakyushu, 1-1 Hibikino, Wakamatsu-ku, Kitakyushu, Fukuoka 808-0135, Japan.

Michael Neugebauer Physikalisch-Technische Bundesanstalt, 38116 Braunschweig, Germany.

Ulrich Neuschaefer-Rube Physikalisch-Technische Bundesanstalt, 38116 Braunschweig, Germany.

Takao Sajima Department of Mechanical Engineering, Faculty of Engineering, Kyushu University, 744, Motooka, Nishi-ku, Fukuoka 819-0395, Japan.

Dong-Yea Sheu Graduate Institute of Manufacturing Technology, National Taipei University of Technology, Taipei 10608, Taiwan.

Yuki Shimizu Department of Finemechanics, Tohoku University, Sendai 980-8579, Japan.

Martin Stein Physikalisch-Technische Bundesanstalt, 38116 Braunschweig, Germany.

Jing-Qi Sun School of Instrument Science and Opto-electric Engineering, Hefei University of Technology, Hefei 230009, China.

Yin Tung Albert Sun Graduate Institute of Manufacturing Technology, National Taipei University of Technology, Taipei 10608, Taiwan.

Kazuhiko Takahashi Engineering Department, MMC RYOTEC Corporation, Gifu 503-2301, Japan.

Rudolf Thalmann Federal Institute of Metrology METAS, 3003 Bern-Wabern, Switzerland.

Kuo-Yu Tseng Micro Machining Laboratory, National Taipei University of Technology, Taipei 10608, Taiwan.

Chenchen Wang School of Instrument Science and Opto-electric Engineering, Hefei University of Technology, No.193, Tunxi Road, Hefei 230009, China.

Chun-Chieh Wang Mechanical and Systems Research Laboratories, Industrial Technology Research Institute, Hsinchu 31040, Taiwan.

Shih-Ming Wang Department of Mechanical Engineering, Chung Yuan Christian University, Taoyuan 32023, Taiwan.

Kui Wu School of Instrument Science and Opto-electric Engineering, Hefei University of Technology, No.193, Tunxi Road, Hefei 230009, China.

Meng Xiang School of Instrument Science and Opto-electric Engineering, Hefei University of Technology, Hefei 230009, China.

Chin-Cheng Yeh Department of Mechanical Engineering, Chung Yuan Christian University, Taoyuan 32023, Taiwan.

Lian-Dong Yu School of Instrument Science and Opto-electric Engineering, Hefei University of Technology, Hefei 230009, China.

Hui-Ning Zhao School of Instrument Science and Opto-electric Engineering, Hefei University of Technology, Hefei 230009, China.

Bin Zhou School of Instrument Science and Opto-electric Engineering, Hefei University of Technology, Hefei 230009, China; Anhui Electrical Engineering Professional Technique College, Hefei 230051, China.

About the Guest Editor

Kuang-Chao Fan is a professor at the Dalian University of Technology in China. He was previously a professor of mechanical engineering at the National Taiwan University since 1989 and a Cheng Kong Scholar at Hefei University of Technology in China since 2001. He is a Fellow of SME and ISNM. His research interests include manufacturing metrology, precision machining, machine tool technology, and micro/nano measurements. He has published more than 400 academic papers and received a number of honors and awards. He was the President of ASPEN from 2012 to 2013.

Preface to "Design and Applications of Coordinate Measuring Machines"

Coordinate measuring machines (CMMs) have been conventionally used in industry for 3-dimensional and form-error measurements of macro parts for many years. Ever since the first CMM, developed by Ferranti Co. in the late 1950s, they have been regarded as versatile measuring equipment, yet many CMMs on the market still have inherent systematic errors due to the violation of the Abbe Principle in its design. Current CMMs are only suitable for part tolerance above 10 μm. With the rapid advent of ultraprecision technology, multi-axis machining, and micro/nanotechnology over the past twenty years, new types of ultraprecision and micro/nao-CMMs are urgently needed in all aspects of society.

This Special Issue collates 11 papers accepted after the review process. These papers present recent advances in coordinate measuring machines, including a new probe design, new machine design, measurement methods, in-process on-machine measurement, uncertainty analysis and state-of-the-art reviews. It is therefore valuable to commercial sectors, research engineers, research students and academics.

I am particularly grateful to all of the contributors without them this Special Issue would not have become a reality. As the guest editor, I wish to acknowledge all the reviewers for their careful evaluation and valuable suggestions to the contributing papers. Special thanks also go to the publishing team of the *Applied Sciences* journal.

Kuang-Chao Fan
Guest Editor

Overview of 3D Micro- and Nanocoordinate Metrology at PTB

Gaoliang Dai, Michael Neugebauer, Martin Stein, Sebastian Bütefisch and Ulrich Neuschaefer-Rube

Abstract: Improved metrological capabilities for three-dimensional (3D) measurements of various complex micro- and nanoparts are increasingly in demand. This paper gives an overview of the research activities carried out by the Physikalisch-Technische Bundesanstalt (PTB), the national metrology institute of Germany, to meet this demand. Examples of recent research advances in the development of instrumentation and calibration standards are presented. An ultra-precision nanopositioning and nanomeasuring machine (NMM) has been upgraded with regard to its mirror corner, interferometers and angle sensors, as well as its weight compensation, its electronic controller, its vibration damping stage and its instrument chamber. Its positioning noise has been greatly reduced, e.g., from $1\sigma = 0.52$ nm to $1\sigma = 0.13$ nm for the z-axis. The well-known tactile-optical fibre probe has been further improved with regard to its 3D measurement capability, isotropic probing stiffness and dual-sphere probing styli. A 3D atomic force microscope (AFM) and assembled cantilever probes (ACPs) have been developed which allow full 3D measurements of smaller features with sizes from a few micrometres down to tens of nanometres. In addition, several measurement standards for force, geometry, contour and microgear measurements have been introduced. A type of geometry calibration artefact, referred to as the "3D Aztec artefact", has been developed which applies wet-etched micro-pyramidal marks for defining reference coordinates in 3D space. Compared to conventional calibration artefacts, it has advantages such as a good surface quality, a well-defined geometry and cost-effective manufacturing. A task-specific micro-contour calibration standard has been further developed for ensuring the traceability of, e.g., high-precision optical measurements at microgeometries. A workpiece-like microgear standard embodying different gear geometries (modules ranging from 0.1 mm to 1 mm) has also been developed at the Physikalisch-Technische Bundesanstalt.

Reprinted from *Appl. Sci.* Cite as: Dai, G.; Neugebauer, M.; Stein, M.; Bütefisch, S.; Neuschaefer-Rube, U. Overview of 3D Micro- and Nanocoordinate Metrology at PTB. *Appl. Sci.* **2016**, *6*, 257.

1. Introduction

Following the progressive miniaturization of today's manufacturing processes, more and more micro- and nanoparts with a complex geometry are applied to

1

numerous industrial products, such as those in the automotive, medical, robotics and telecommunications fields. Full 3D measurements of these micro- and nanoparts with uncertainties down to 100 nm or even below are increasingly in demand [1,2]. For instance, spray holes of fuel injection nozzles are to be fabricated with diameters of less than 100 µm for a better fuel atomization. Measurements of the diameters and of the form and inner surface quality of the spray holes are of crucial importance. The microgears with modules from 1 µm to 1 mm are key components of, e.g., microrobotics and medical devices. Nondestructive measurements and the quality control of both the mould and the replicated gears are of great importance.

Today various techniques are available for full 3D measurements of microparts. One of the most important development trends in industrial dimensional metrology, having the potential to fulfil the requirements to measure complex microparts, is multi-sensor coordinate metrology. This combines the speed of optical measurements with the accuracy and 3D capability of tactile measurements and, more recently, the ability to measure interior features using X-ray computed tomography (CT) [3]. Improved metrological capabilities are needed to ensure the measurement traceability and reliability of the various measurement techniques.

To offer the highly accurate full 3D metrological capability of microparts, a generation of micro-coordinate measuring machines (micro-CMMs) has been developed in the last two decades [4–14]. The first micro-CMM was developed by Peggs et al. [4] at the National Physical Laboratory of the United Kingdom in the year 1999. In its configuration, they applied a mirror corner near the CMM probe as reference mirrors and utilized three laser interferometers and three autocollimators to measure the probe position with respect to the metrology frame. This novel design greatly reduced the Abbe offset, thus offering high 3D measurement accuracy (estimated as 50 nm at the 95% confidence level) over a measurement volume of 50 mm × 50 mm × 50 mm. Almost at the same time, Vermeulen et al. [5] developed a micro-CMM where linear scales are applied to measure the position of the probe tip fully in compliance with the Abbe principle in the x- and y-directions with a motion volume of 100 mm × 100 mm × 50 mm. The design was later commercialized by the Zeiss company in their micro-CMM F25 (unfortunately, F25 is now no longer in the product portfolio of Zeiss). A similar design idea has recently been realized in the micro-CMM "TriNano" with all the three axes measured with the Abbe principle [6]. In the year 2000, Jäger et al. [7] developed a nanopositioning and nanomeasuring machine (NMM) with a motion volume of 25 mm × 25 mm × 5 mm. They applied three miniature laser interferometers and two autocollimators to measure six degrees of freedom (DOFs) of the motion stage with respect to the Zerodur metrology frame, which is fully in compliance with the Abbe principle in all three axes. Using a similar principle, Ruijl [8] built the CMM "Isara" with a measurement volume of 100 mm × 100 mm × 40 mm. Recently, a larger version ("Isara 400") has been

developed by IBS Precision Engineering BV with a motion volume of 400 mm ×
400 mm × 100 mm [9].

Various micro-CMM probes have also been developed [4,11,15–30]. Most
of them are mechanical tactile probes working in the contact mode [4,11,15–19].
In their configuration, typically a stylus (having a probing sphere at its free end,
and usually fixed to a rigid centre plate/boss at the other end) is suspended by
a type of flexure structure, for instance, flexure strips [4,15], slender rods [14,18],
flexure hinges [11] or membranes [16]. When the probing sphere is touched and
deflected by a workpiece, the flexure structure is deformed due to the probing force.
By measuring the deformation via various sensing techniques, for example by means
of capacitive sensors [4,15], piezoresistive sensors [16], inductive sensors [11], laser
focus sensors [14], a Michelson interferometer combined with an autocollimator [17],
or even by means of low-cost DVD pick-up heads [18], the translational motion
of the probe sphere in 3D can be derived. Some of the probes [15,16] are realized
based on the micro-electro-mechanical systems (MEMS) technique, which offers
a much smaller dynamic mass (20 mg) and thus enables a potentially higher
measurement speed.

Although being the mostly used probing technique in micro-CMMs and offering
outstanding measurement performance (3D uncertainty typically in the order of tens
of nanometres), tactile probes have several limiting aspects. The first aspect concerns
the probing force and the probing stiffness. A design trade-off needs to be made
between the stiffness and the flexibility of the probing element, for instance to avoid
damaging the sensitive parts on the one hand, and to overcome the surface forces
or inertial loads on the other hand. To overcome this trade-off, the University of
Nottingham (UK) has recently developed a probe whose stiffness is variable due to
the use of a switchable suspension structure [19]. The second limitation is related
to the contact measurement mode. This method has several drawbacks—such as
possible surface damage, unwanted adhesion to surfaces, or measurement bias due
to the inertia force. In addition, the contact mode microprobes usually apply styli
with a limited aspect ratio to achieve a better measurement stability. To mitigate
these problems, non-contact mode micro-CMM probes have been developed [21–23].
In the non-contact measurement mode, the probes are vibrated and the change of
the vibration amplitude due to the probe sample's interaction is usually applied
for measurements. For instance, Bauza et al. [21] developed a high-aspect-ratio
microprobe using a probe shank with a diameter of 7 μm having an aspect ratio of
700:1. During the measurements, its probe shank is oscillated in a standing wave.
The motion of the free end of the probe shank forms a virtual probe tip to serve as
the contact point, thus no spherical ball is required. Claverley et al. [22] developed
a novel vibrating tactile probe for which six piezoelectric actuators and sensors are
fabricated using electro-discharging machining on the three legs of its triskelion

design. However, the shape and stability of the motion trajectory of the non-contact probing tip are a critical issue. Murakami et al. [23] reported that the shaft moved in an elliptical motion in their design. Therefore, a compensation method is applied by measuring the stylus' displacement using position sensitive detectors (PSDs). However, the PSDs are bulky and consequently limit the available measurement space. The third limitation concerns the measurement loop. The measuring sensors of most of the tactile probes are typically located at the suspending membrane or at the flexure strips and measure the signal, which is being transferred via the probing styli. Consequently, the deformation of the probing styli is "invisible" in the sensor's readout, which leads to measurement errors. Such a problem becomes much more serious for probes with a smaller styli diameter (d) and a longer styli length (l) as the deformation is inversely proportional to d^4 and proportional to l^3 [20]. To tackle these problems, a tactile optical probe—well known as the fibre probe today—has been developed at the Physikalisch-Technische Bundesanstalt (PTB) [24]. In contrast to the tactile probes, the measurement signal in fibre probe (i.e., the position of the probing sphere) is directly detected by an optical CCD camera without the mechanical transfer of the probe styli [24]. The advantages are that the probing forces can be very low (1 μN to 100 μN) and stylus tips are available with much smaller diameters (down to 25 μm). The 2D version fibre probe has been used for many years to measure injection nozzles, turbine blade cooling holes, and a variety of other small, tight tolerance features. An upgrade of this probe, the 3D fibre probe, has been commercially available for some time [25]. PTB collaborated with the Werth Messtechnik company are continuing the investigations on this microprobe, which will be detailed in this paper. The design idea of the fibre probe was followed by Cui et al. [26] who proposed a spherical coupling fibre probe as an attempt to overcome the shadowing effect. Some other kinds of tactile-fibre probes have also been proposed, where Bragg grating strain sensors [27], fibre optical displacement sensors [28] and micro-focal-length collimation techniques [29] are applied to measure the position of the probing sphere or of the stylus. In addition, Weckenmann et al. [30] proposed an electrical tunnelling current probe for force-free probing; and Michihata et al. [31] put forward a probe based on the laser trapping technique.

System calibration and performance verification are crucial tasks for 3D micro-coordinate metrology. To qualify, calibrate and verify these micro-CMMs, calibration standards as well as the traceable calibration of these standards are essential. However, this is still a significant challenge today due to the extremely low uncertainty demanded, as is also summarized in a review paper by Claverley et al. [32]. It is extremely difficult to achieve the required calibration uncertainties at calibrated test lengths, due to the limited availability of both high-quality physical standards and metrological services. For example, with the current micro-CMMs exhibiting an uncertainty of 100 nm or less, it is essential for any test length used for verification to be calibrated with an uncertainty that is five- or even ten-times lower, i.e., an uncertainty of better than 20 nm. The micro-CMMs are not covered by the current international standard documents for CMM testing or are only partly covered by them (e.g., VDI 2617 12.1 [33]). Problems also occur when trying to apply the existing acceptance tests defined in ISO 10360-2 [34] to micro-CMMs. According to the test procedures defined in ISO 10360-2, for example, it is required to measure five different calibrated test lengths which are located at seven orientations within the measurement volume of the CMM, four of which must be the space diagonals. However, the shaft length of a micro-CMM probe is usually kept short in order to enhance the measurement stability, which limits its measurement accessibility to all reference features of the calibration artefact. Therefore, the development of a traceable metrological capability and of physical as well as documentary standards is still an urgent task for promoting the commercialization and application of micro-coordinate measuring tools today.

State-of-the-art tactile micro-coordinate metrology has recently been reviewed by Thalmann et al. [35] and three key aspects—stage and metrology system design, probe developments, and system calibration and performance verification—have been well summarized. Therefore, this paper is focused on giving an overview of the research activities carried out at PTB.

2. Instrumentation Developments

2.1. Upgrade of a Nanomeasuring and Nanopositioning Machine (NMM)

Several micro/nano-CMMs are operated at PTB, including a SIOS NMM, a Zeiss F25, and a Werth VideoCheck UA CMM. Here, we detail the recent upgrade of the NMM in collaboration with the SIOS Company, Ilmenau, Germany.

The measurement principle of the NMM is briefly shown in Figure 1a. Its motion platform consists of a mirror corner which comprises three high-precision planar mirrors attached orthogonally to each other. With three high-precision interferometers (x-, y- and z-interferometer), the displacement of the motion platform can be measured with respect to the metrology frame (Zerodur frame) with a resolution of 0.08 nm. In addition, there are two angle sensors available for measuring all three angular DOFs of the motion platform with a resolution of better than 0.01″. Thus, all six DOFs of the motion platform are accurately measured. The motion platform is moved by three stacked mechanical stages driven by voice coil actuators (not shown). By utilizing a digital signal processor (DSP) servo controller based on the measured six DOF values, the NMM is capable of positioning and measuring with nm accuracy. For micro/nano-CMM measurements, the sample is fixed on the mirror corner and the CMM probe is typically located at the intersection point of three measurement beams of laser interferometers. Thus, the measurement is performed fully in compliance with the Abbe principle along all three axes.

The recent upgrade of the NMM was undertaken with regard to a number of components, as detailed below:

- The geometry of the corner mirror has been improved to fix samples better so that the stress introduced into the optical component due to the sample fixing is greatly minimized. In addition, the height of the mirrors has been increased so as to allow higher objects (up to 22 mm).
- The interferometers and the angle sensors have been improved for easier adjustment, better thermal behaviour and better stability.
- A new motorized spring mechanism has been installed for the weight compensation of the motion stage. As a result, the heat generation of the z-driving motors is greatly reduced, allowing much better temperature stabilization.
- All the control electronics have been upgraded. They now have an increased servo frequency response up to 1 kHz for better stage control performance.
- An improved instrument chamber for better thermal and acoustic insulation and an improved vibration damping stage are applied.

To demonstrate how much the performance of the NMM has improved, the positioning noise along the z-axis before and after the upgrade, measured at the same sampling frequency of 6.25 kHz, is compared in Figure 1b. It can be seen that the noise level has been significantly reduced from $1\sigma = 0.52$ nm to $1\sigma = 0.13$ nm. In Figure 1c, a positioning example is shown where the NMM is commanded to move along the x-, y- and z-axes simultaneously by several steps of 10 μm with a speed of 5 μm/s. The position noise along the x-, y- and z-directions after arriving at the target positions is shown in Figure 1d, indicating an excellent positioning and measurement performance.

2.2. Boss-Membrane Piezoresistive Microprobe

Several micro-CMM probes are being further developed at PTB, including a boss-membrane piezoresistive microprobe, a fibre probe and probes based on atomic force microscope (AFM).

Figure 2 shows the measurement principle of the boss-membrane piezoresistive microprobe. The fabricated sensor chip includes a centre boss, a membrane having a thickness of tens of micrometres, and a frame. On this chip, a shaft with a length of about 10 mm is glued to the centre boss, and a probing sphere with a diameter of some hundred micrometres is glued to the free end of the shaft. Four groups of piezoresistive sensors, arranged as Wheatstone bridges as shown in Figure 2b, are fabricated on the back of the membrane by the ion implantation technique and act as sensor elements. When the probing sphere touches the measurement object, strains are produced on the membrane by the probing force which leads to changes in the resistances of the piezoresistive sensors. The finite element method (FEM) has been used during the probe design to calculate the position where the maximum strains occur. At these positions, the piezoresistive sensors are located in order to achieve optimum measurement sensitivity. The resistance changes of the sensors are converted into electric signals which are used to determine the probe's displacement, i.e., for measurement. A photo of such a sensor chip is shown in Figure 2c. The probe was originally designed and fabricated at the Institute for Microtechnology of the Technical University Braunschweig (Braunschweig, Germany) [16], and PTB has applied this technique to micro-CMM applications and has fully investigated its performance in order to achieve improvements [10,36].

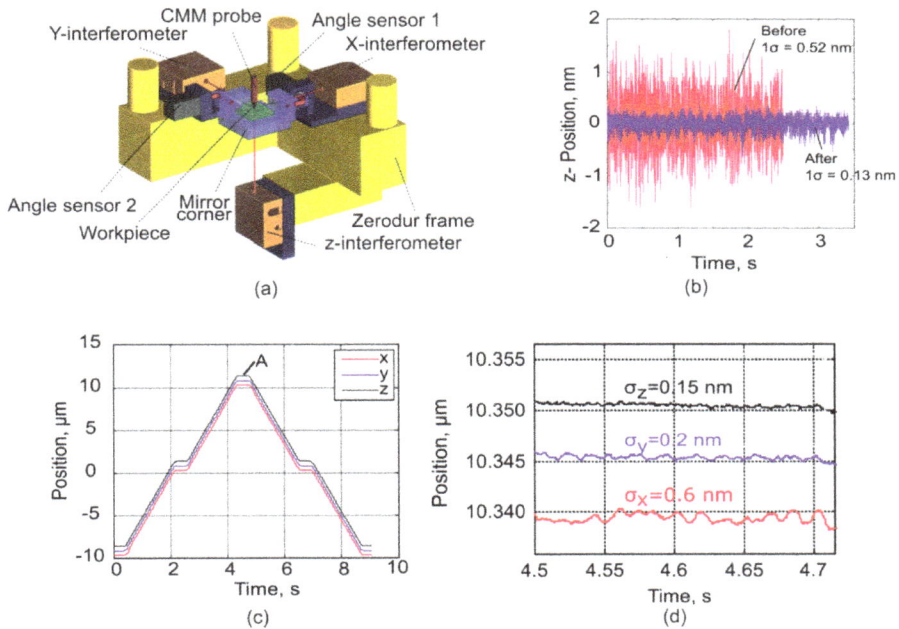

Figure 1. (**a**) Schematic diagram showing the measurement principle of the nanopositioning and nanomeasuring machine (NMM); (**b**) positioning noise along the z-axis before and after machine upgrade; (**c**) example showing the positioning of the NMM along the x-, y- and z-axes simultaneously by steps of 10 μm; and (**d**) the positioning stability of three axes after reaching the target position.

A major shortcoming of the probe is its anisotropic stiffness. For instance, for one probe which was investigated in detail, the stiffness values along the x-, y- and z-axes were 208.8 N/m, 313.8 N/m and 5642.9 N/m, respectively. As the styli deformation differs in the different probing directions, such behaviour will lead to form measurement errors, particularly in the scanning measurement mode. In addition, it may also result in a slipping of the probing sphere with respect to the workpiece during the measurements.

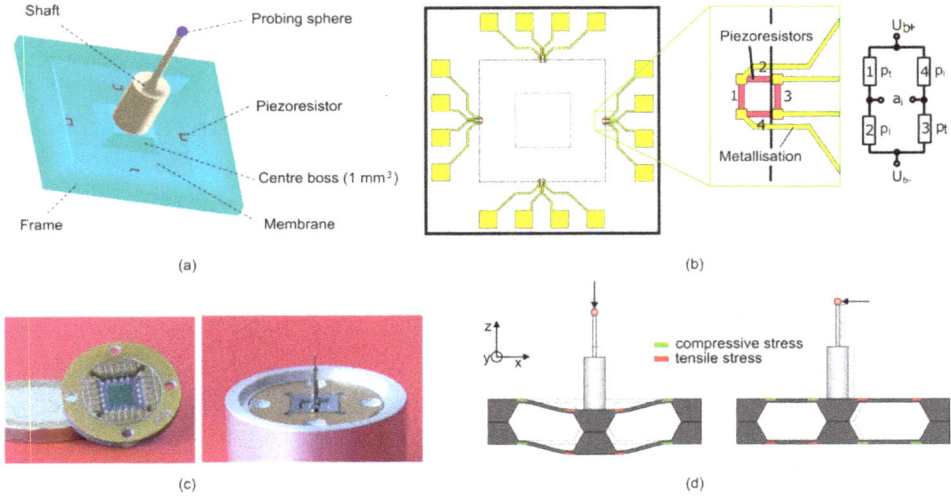

Figure 2. (a) 3D structure of the piezoresistive micro probe; (b) layout of the piezoresistance sensors (Wheatstone bridges) at the back of the membrane; (c) photos of the fabricated sensor chip and of the micro-3D-CMM probe as a whole. A one euro coin illustrates the size of the probe; and (d) a double triangle design realized for improving the isotropy of stiffness. Figures (a)–(c) are reproduced with permission from [36], Copyright IOP Publishing Ltd., 2009.

To solve this problem, a double triangle design has recently been realized which consists of two structured boss membranes mounted face-to-face as shown in Figure 2d [37]. The states of stress occurring in this structure when it is loaded with vertical and lateral probing forces are also illustrated in the figure. In early stages of the development, the sandwich structure was fabricated by gluing two single chips together whereby one of the chips did not have any electronic components on it. The capillary forces which occurred during the gluing processes yielded a very good alignment of the two chips. Experimental investigations show that the stiffness ratio between the x-, y- and the z-direction has been improved to approximately 2 by the improved design [37]. Currently, a joint research project is also being carried out by the Technical University of Braunschweig in collaboration with PTB to further develop the probe. With the improved design, the probe is capable of measuring with a probing speed of up to 1 mm/s and of achieving a 3D probing repeatability of 50 nm. Testing its probing error according to ISO 10360 has not been done yet.

2.3. 3D Tactile-Optical (Fibre) Probe

The principle of a 3D fibre probe is illustrated in Figure 3a. The microprobe consists of an optical glass fibre which acts as the stylus, with a small spherical tip attached to the end. The tip is mounted in the focal plane of the imaging system of an optical coordinate measuring machine (CMM). It is mirror-coated on the lower hemisphere in order to achieve reflectivity. The fibre is fixed to an optical CMM by a three-curved prong leaf spring of low stiffness. This arrangement ensures the flexibility of the probe in all axes.

The determination of the tip position in the axes horizontal to the optics (x- and y-directions) is similar to the well-known 2D fibre probe. The image of the illuminated stylus tip is located in the camera image of the optical CMM by correlation techniques. The stylus position along the z-axis is determined by an optical distance sensor, e.g., based on the Foucault knife-edge principle [38], which measures the height of the upper end of the fibre.

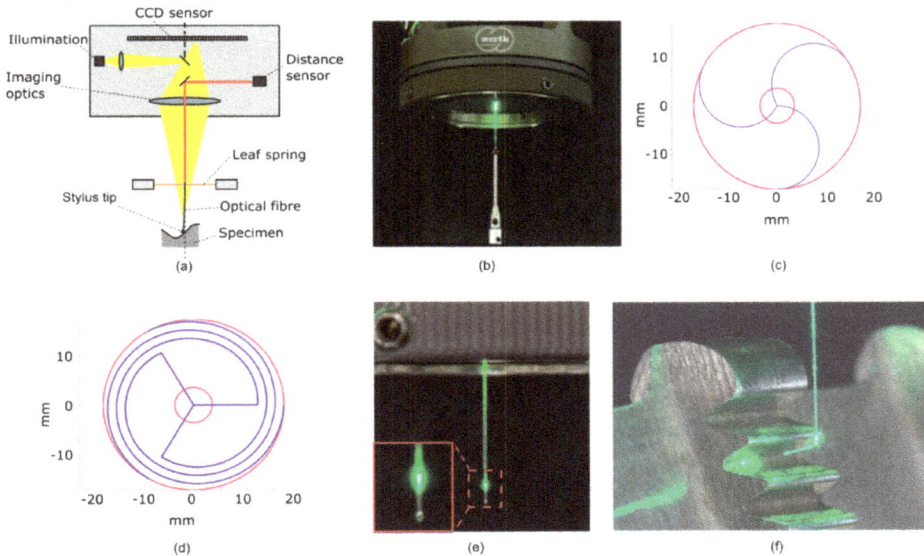

Figure 3. (**a**) Schematic diagram showing the measurement principle of a 3D fibre probe; (**b**) photo of a 3D fibre probe measuring a ruby sphere (Φ 2 mm); (**c,d**) the designs of the leaf spring before and after improvement, respectively; (**e**) an alternative design with a dual-sphere stylus; and (**f**) an alternative design with an L-shaped stylus [25].

Apart from the 3D measurement function, also the design of the fibre probe has been further improved [25]. For instance, in order to achieve a better isotropic probing stiffness, the design of the leaf spring which suspends the stylus has been optimized by FEM calculations. The original and the optimized designs are shown in Figure 3c,d, respectively. With this improved design, the ratio between the stiffness values in vertical and in horizontal direction could be reduced down to 1.4:1 for standard probes (tip diameter: approximately 250 μm). Furthermore, to eliminate any shadowing effects by the workpiece, a stylus with two spheres has been developed as shown in Figure 3e,f. In such a configuration, the lower sphere is applied to probe the workpiece, whereas the upper sphere is applied for measurements. The displacement ratio between the two spheres was calibrated prior to use. As an advantage, the imaging of the upper sphere is not limited by the workpiece and therefore no shadowing effect occurs. The simplest realization—which consists of a stylus with both spheres arranged vertically with one above the other—is shown in Figure 3e. The distance between the two spheres can be up to a few millimetres. This allows measurements in holes, without the upper sphere being obstructed by the sidewalls of the hole, and thus without reducing the measurement accuracy by this optical effect. However, it is to be mentioned that larger distances between the two spheres lead to larger measurement deviations due to the smaller displacement of the upper sphere. Therefore, the distance between the spheres should be as small as possible, depending on the measurement application.

Another advanced design of the dual-sphere stylus is the L-shaped configuration, as shown in Figure 3f. The stylus consists of a single fibre with two spheres, whereby the fibre is bent below the upper sphere. With such a probe, structures with undercuts can be measured. Here, the bent angle and the length of the bent part can also be adapted to the measurement task.

The 3D fibre probe is capable of achieving a typical probing speed of 0.1 mm/s to 3 mm/s. The specified probing errors are:

- Single point probing (ISO 10360-5): $P = 0.25$ μm (probe diameter 250 μm), $P = 0.5$ μm (probe diameter 40 μm and 100 μm); and
- Scanning (ISO 10360-4): THN = 1.5 μm (probe diameter 250 μm), THN = 2 μm (probe diameter 40 μm and 100 μm).

2.4. AFM-Based 3D Probes

Although the diameter of the smallest micro-CMM probe may be as small as 25 μm, it is still too large to measure 3D structures with sizes of a few micrometres or even below. There is a metrology gap between AFMs—the most popular used coordinate measuring techniques for nanostructures—and micro-CMMs. To fill this gap, another idea is a type of 3D probe based on the AFM technique. At PTB, we have developed a 3D-AFM [39] and the so-called assembled cantilever probe [40], which is promising to fill this gap.

The measurement principle of the 3D-AFM developed by PTB is shown in Figure 4a. It utilizes flared AFM tips. Such tips have an extended geometry near their free end which enables the probing of steep and even undercut sidewalls. The probe element can be regarded as a disc; however, due to its tiny size, it still has a high spatial resolution. Similar to conventional AFMs, the 3D-AFM can measure in contact, intermittent and non-contact mode. Furthermore, a vector approach probing (VAP) method has been applied for enhancing the measurement flexibility and for reducing tip wear [24]. Our preliminary uncertainty estimation indicates that the 3D-AFM is capable of measuring the feature width of nanostructures with an expanded uncertainty down to 1.6 nm (at a confidence level of 95%).

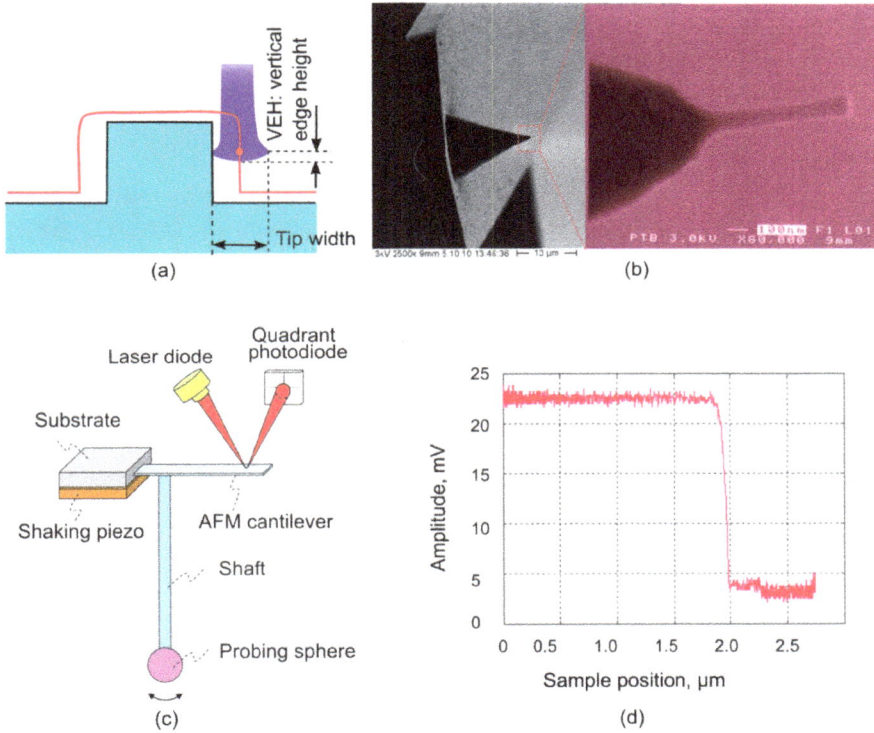

Figure 4. (a) Schematic diagram showing the measurement principle of a 3D-AFM probe; (b) SEM image taken of a flared AFM tip applied in a 3D-AFM; (c) schematic diagram of 3D assembled cantilever probe (ACP); and (d) a typical probing curve of the 3D-ACP probe. Figures (c), (d) are reproduced with permission from [40], Copyright American Institute of Physics, 2007.

Currently, such flared probes are commercially available with a diameter up to 850 nm and an effective stylus length of 7.5 μm [41]. However, the technology can be expanded by building larger flared tips with diameters of a few micrometres and styli lengths of tens of micrometres using, e.g., Focused Ion Beam (FIB) milling or FIB induced deposition techniques.

However, due to the limited speed of the FIB microfabrication technique, the size of its manufacturable probing elements is still limited. To further expand the AFM technique for full 3D measurements, a type of assembled cantilever probe (ACP) has also been proposed [40]. Compared to the flared tip where the probing element is directly fabricated on the tip, the ACP technique applies microassembling techniques to create a probing stylus. The ACP also has the advantages of the AFM technique, such as high measurement sensitivity, very low measurement forces (μN to nN level), a compact structure, low fabrication costs and the ease of probe exchanges. Additionally, the ACP probe is mechanically and electrically compatible with commercial AFMs. It could therefore be applied directly in commercial AFMs to extend their functions from surface topography measurements to 3D measurements, if some minor software modifications are provided.

Due to its low probing stiffness (typically about 1.5 N/m in lateral directions), the 3D-ACP probe is usually used for point-to-point measurements. It is capable of performing measurements with a probing speed of up to 100 μm/s. It has a probing repeatability of approximately 25 nm (p-v) along the z-axis, and of approximately 130 nm (p-v) in full 3D, as shown in Figure 5. Testing of its probing error according to ISO 10360 has not been performed yet.

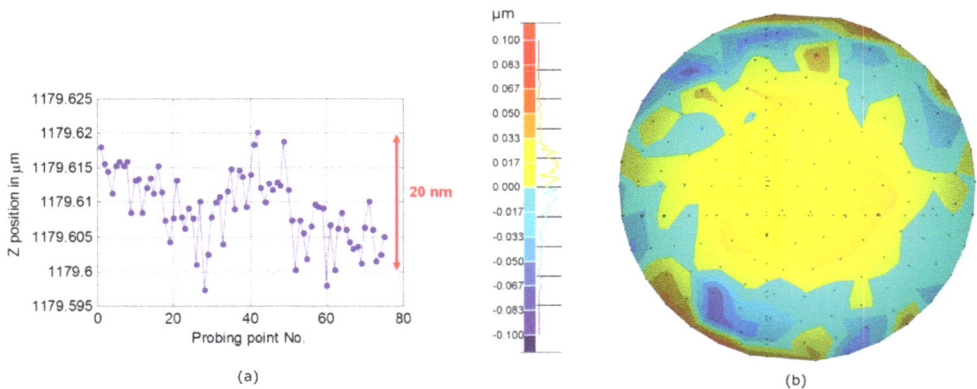

Figure 5. (**a**) Probing repeatability of a 3D-ACP probe along the z-axis; and (**b**) probing repeatability of a 3D-ACP probe in full 3D.

3. Calibration Artefacts

Calibration artefacts are essential for acceptance tests, as well as for the calibration and verification of micro-CMMs. Several categories of calibration artefacts have been developed at PTB for calibrations of, for instance, the probing force and stiffness, geometric errors, contour and microgear measurements. This section introduces some state-of-the-art calibration artefacts developed at PTB.

3.1. Microforce Calibration Standard

To calibrate the probe's force and stiffness, a convenient method is to apply a reference spring, whose spring constant has been calibrated by, e.g., means of a compensation balance before usage [42]. An application example was introduced in [36]; however, it is only suitable for calibrating the z probing axis due to its size limit.

To overcome this problem, a new artefact has been fabricated as shown in Figure 6. The artefact consists of three cantilevers marked as "x", "y" and "z" used as reference springs, whose spring constant has been calibrated *in prior*. The cantilevers and their substrates are made from silicon using etching techniques. The substrates are glued to an aluminium cube with a glass plate (thickness: 0.2 mm) inserted as a spacer. This glass spacer ensures that the cantilever will not contact the aluminium cube in its free standing and bent states. Near the free end of the cantilever, a trapezoid-shaped tip is fabricated. It is used for loading the probing forces at a known position. As the spring constant of the reference cantilever depends on the position where the force is loaded, the cantilever should be calibrated and used for calibration with forces loaded at the given position. Using such a device, we calibrated the x-, y- and z-stiffness of a boss-membrane 3D probe to be 208.8 N/m, 313.8 N/m and 5642.9 N/m, respectively, as mentioned earlier.

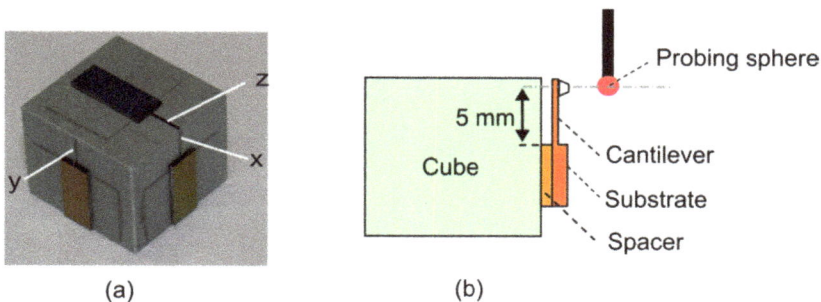

(a)   (b)

Figure 6. (a) Photo of a cube with three cantilevers used for calibrating the x, y and z spring constants of micro/nano-CMM probes; and **(b)** structure of a cantilever for probing force calibration.

Adhesive forces are observed during the probing process resulting in sticking between the probing sphere and the sample [20]. Such adhesive forces may be attributed to several factors such as atomic/molecular interaction forces, capillary forces, electrostatic forces etc. Since the probing forces of micro/nano-CMMs are only in the order of milli- or micronewtons, the adhesive forces may impact the CMM's measuring/scanning performance significantly.

Such adhesive forces can also be characterized using the force calibration standard. Figure 7 shows an example where the adhesive force of an ACP probe is characterized. Figure 7a shows a typical probing curve, which depicts the cantilever bending signal vs. the probing distance. For ease of explanation, the curve is marked with letters "A" to "F" to indicate different probe-sample interaction states as shown in Figure 7b. B is the point where the probing sphere begins to contact the sample. This point is calculated as the zero force contact point. In B → C, the probe moves towards the sample. The probing force leads to a positive change of the probing signal, which is quasi-linear with respect to the probing distance. At point C, the motion is stopped and the probing sphere begins to retract from the sample. In C → D, the probing sphere undergoes the reverse trajectory to that of B → C and no hysteresis is found. During D → E, the probing sphere sticks to the sample. The adhesive forces lead to a negative change of the probe signal. At point E, the probing sphere begins to separate from the sample. In E → F, a type of damped oscillation can be seen, which is caused by the release of the adhesive forces on the bent probe. The adhesive forces F_a can be calculated as the probing forces at point E according to Hooke's law:

$$F_a = k_x \cdot L_r \tag{1}$$

Here, k_x is the spring constant of the probe in the x-direction and L_r is the probing distance needed for separating the probing sphere and the sample.

In this study (sample: sapphire sphere with $\varnothing = 2$ mm; probing sphere: sapphire sphere, $\varnothing = 0.12$ mm; relative humidity: $46\% \pm 2\%$; temperature: $20.5 \pm 0.5\ °C$), we measured $L_r = 1.31\ \mu m$ and estimated adhesive forces of 2.06 μN according to the calibrated reference spring constant $k_x = 1.575$ N/m.

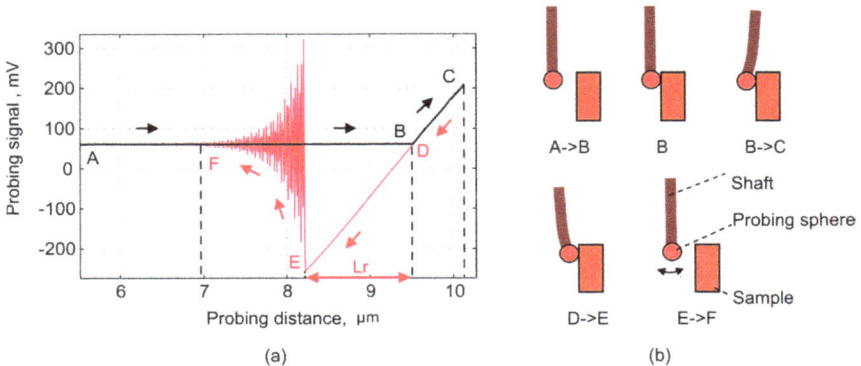

Figure 7. (a) Recorded probing curve for estimating the adhesive forces between probing sphere and sample; and (b) different probe-sample interaction phases.

14

3.2. 3D Aztec Artefact

Geometry calibration standards are essential for the verification or calibration of geometric errors of micro-CMMs. A number of standards have been developed worldwide. Most of them look like a miniature version of popular artefacts applied in large CMMs, for instance, mini ball bar [43], mini ball array [44] or mini ball plate. Photos of some representative calibration artefacts studied at PTB are illustrated in Figure 8.

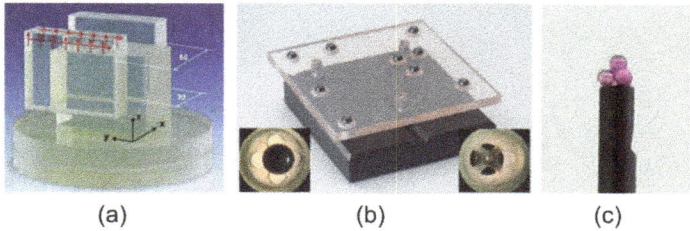

(a)   (b)   (c)

Figure 8. Photos of several geometry calibration standards for micro-CMMs, shown as: (**a**) a Zerodur gauge block bridge; (**b**) a ball plate with nine hemispheres made of silicon nitride manufactured by Carl Zeiss IMT; and (**c**) a micro-tetrahedron artefact consisting of four spheres with (Ø 0.5 mm).

However, there is a practical limit in applying the artefacts mentioned above. The surface area of reference planes/spheres measurable by microprobes is limited by either the structure of the artefacts or the styli length, consequently impacting the calibration accuracy. In addition, the precise assembling of tiny reference spheres on substrates is also not a trivial task.

To mitigate the problems mentioned above, recently a type of 3D Aztec artefact manufactured from crystal silicon using the wet-etching technique has been developed. One basic unit of the Aztec artefact is shown in Figure 9a. It has a pyramidal shape and consists of a few plateaus. On each plateau, micro-pyramidal marks are fabricated, as detailed in Figure 9b. To calibrate the marks, usually four sidewall planes of marks are measured and their intersection point is calculated as the reference coordinate. In the given design example, the overall size of one pyramid artefact is 6.5 mm × 6.5 mm × 1.4 mm. A photo of a fabricated Aztec artefact in 4-inch wafer size is shown in Figure 9c. It can be conveniently sawed into a suitable size to fit the measurement volume of different micro-CMMs.

The Aztec artefact has several advantages. For instance, the artefact can be mass-produced using the wet-etching technique cost-effectively; due to the crystal nature of the silicon material, the marks have high sidewall surface quality and well-defined geometry (angle between sidewalls and upper surface: 54.7°); the layout of the artefact has a wide open space from the top, offering better accessibility for the probing styli.

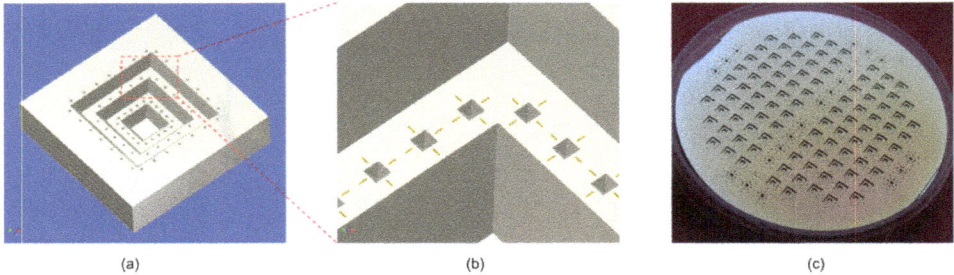

Figure 9. (a) Layout of a 3D Aztec artefact for calibrating micro/nano-CMM; (b) detail of (a); and (c) photo of a fabricated artefact in four-inch wafer size using the photolithography and micromachining technique.

Figure 10 shows the repeatability of a calibration of the 3D Aztec artefact using the NMM equipped with a boss-membrane probe. Two repeated measurements are run, with each offering a set of reference coordinates. Their coordinate difference is shown as vectors for the x- and y-coordinates in Figure 10a and that for the z-coordinate in Figure 10b. The scale of the vectors is shown in the bottom left region of the plot. It can be seen that for most measurements, the deviation is less than 10 nm, indicating excellent measurement repeatability.

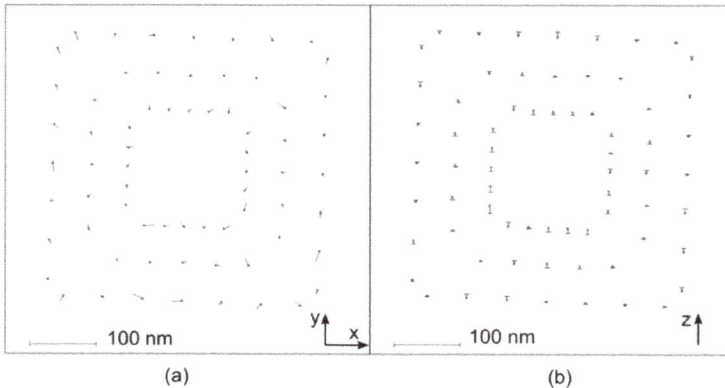

Figure 10. Measurement repeatability of mark coordinates measured on a 3D Aztec artefact by the PTB NMM equipped with a boss-membrane probe. The artefact is calibrated in two repeated measurement runs. The difference of the results is shown as vectors: for the x- and y-coordinates (a); and for the z-axis (b). The scale of the vector is shown in the bottom left region of the plot.

16

3.3. Micro-Contour Standard

To ensure traceability, especially for high-precision optical measurements at microgeometries, a task-specific micro-contour standard was developed in cooperation between the Alicona Company, IPK-Fraunhofer and PTB. This standard is a further development of a standard presented in [45] and, in contrast to the former standard, is calibrated by tactile probing. The micro-contour standard is made of tungsten carbide, is manufactured by wire-EDM at IPK-Fraunhofer and has a diffusely reflecting surface of $Rz \approx 1.5$ μm. It is, therefore, well suited for optical measuring tools such as confocal and focus variation instruments. The external size of the standard is 47 mm × 15 mm × 3 mm with different spherical and prismatic geometric elements on top, as shown in Figure 11a. These elements have dimensions of 0.05 mm up to 5 mm and represent different measurands like radii, angles and step heights. The form deviations of the geometrical elements are in the range of a few tenths of a micrometre.

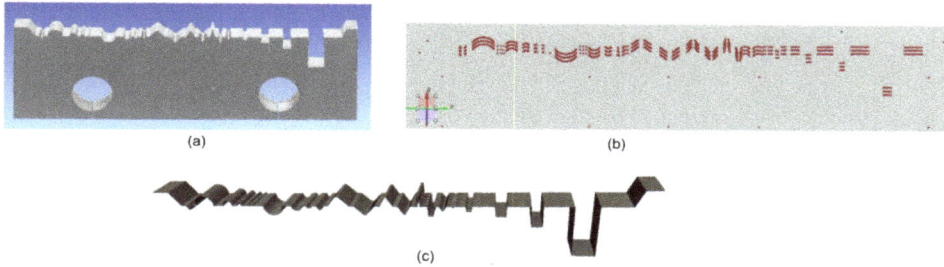

Figure 11. (a) Micro-contour standard having a size of 47 mm × 15 mm × 3 mm; **(b)** 3D data set of the standard obtained with micro-CMM F25, overall 10,800 points; and **(c)** 3D data set of the standard measured with a InfiniteFocus (5× objective) system.

The micro-contour standards are calibrated by tactile single-point probing, usually as working standards at IPK-Fraunhofer, currently using a Zeiss CMM O-Inspect with a probing sphere of Ø 300 μm. The uncertainties amount to about 1.2 μm for radii, 0.2° for angles and 0.6 μm for step heights. For some special purposes which require lower uncertainty, the standards are calibrated as reference standards at PTB. For this, a Zeiss micro-CMM F25 is used with a probing sphere of Ø 120 μm and very low contact forces of approximately 1 mN. The probing deviation of the F25 according to ISO 10360 determined at different reference spheres of Ø 1 mm up to Ø 10 mm amounts to $PF < 0.15$ μm (form error) and $PS < 0.07$ μm (size error). The standards are measured in a horizontal position and the geometric elements are probed in three different traces: 1 mm, 1.5 mm and 2 mm below the front face. Each geometric element in each trace is measured with 50 points, which result in about 10,800 points overall, as shown in Figure 11b. The calibration results are determined from the 3D geometrical elements with 3×50 points each. Figure 10c shows an overview scan result of the standard obtained with an optical focus variation instrument (Alicona InfiniteFocus) with $5\times$ magnification. For the testing of optical instruments, the magnification used is adapted to the size of the geometrical element to be measured.

To test the suitability of using the standard, comparison measurements were carried out between the InfiniteFocus and the F25 at four different standards. The results agree with respect to the measurement uncertainty stated as summarized in Table 1.

Table 1. Results of comparison measurements at four micro-contour standards between InfiniteFocus and F25. R_{CV} is the radius convex; R_{CC} stands for the radius concave; A, the angle; H, the step height; U for $k = 2$, Δ_m for the mean of the absolute values of the differences, E_n-value.

	R_{CV} (μm)	R_{CC} (μm)	A (°)	H (μm)
U_{IF}	2.0	2.0	0.15	1.0
U_{F25}	0.8	0.8	0.1	0.5
Δ_m	1.2	0.4	0.05	0.2
E_n	0.6	0.2	0.3	0.2

3.4. Microgear Standard

Microgears with transverse modules between 1 μm and 1 mm have become an indispensable part of modern production [46]. They are used in medical devices, semi-conductor manufacturing, microrobotics and precision engineering and are, thus, increasingly gaining in economic relevance. For these gears, a minimum amount of material and simultaneous a maximum amount of precision and efficiency are required. For the implementation of these requirements, reliable quality assurance and, thus, reliable measurement technology are indispensable. However, suitable micro-measurement standards and comparison concepts, with the aid of which the measurements are reliably traceable to the "metre", the SI unit of length, have been lacking so far.

PTB has recently developed a workpiece-like microgear measurement standard as shown in Figure 12 with detailed design parameters. The standard artefact embodies different gear geometries on one component. It has modules ranging from 0.1 mm to 1 mm, being adapted to the requirements of industry. The design allows measurements with tactile and optical sensors as well as CT procedures.

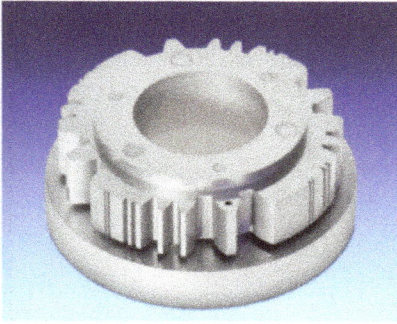

Parameter	Description
Type	external gear
Helix angle β	0°
Addendum modification x	0
Number of teeth z	198 / 99 / 38 / 18
Normal module m	0.1 mm / 0.2 mm / 0.5 mm / 1 mm
Pressure angle α	20°
Tooth depth h	0.2 mm / 0.3 mm / 1 mm / 1.5 mm
Tip diameter d	20 mm
Material	Hard metal or titan

Figure 12. Photo of a workpiece-like microgear measurement standard (tip diameter: 20 mm, module 0.1 to 1 mm) developed at PTB. The design parameters of the standard are shown in the table.

To demonstrate the quality of the fabricated microgear standard, the measured profiles of a spur gear tooth of the standard are shown in Figure 13. It can be seen that the form deviations are smaller than 0.5 μm at both the profile and the helix scans. The microgear measurement standard has been calibrated using the micro-CMM F25 at PTB and was used in a national intercomparison with a broad range of measuring machines involved. For more information, the reader is referred to [47,48].

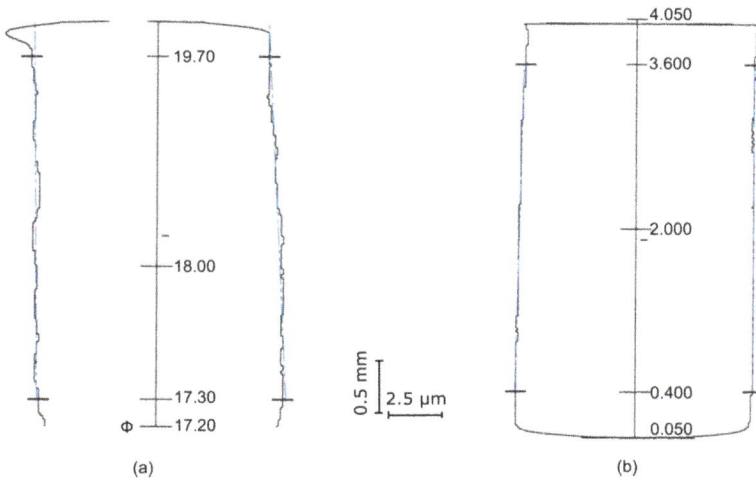

Figure 13. Measured lines on a spur gear tooth (module of 1 mm) of the standard, shown as: (**a**) profile; and (**b**) helix.

4. Conclusions

Micro-coordinate measuring machines (micro-CMMs) are being increasingly applied for accurate 3D measurements of micro/nano parts such as injection nozzles, turbine blades, microholes, and a variety of other small, tight tolerance features. Today, the developments of calibration standards as well as the metrology capabilities are crucially required for acceptance tests, along with the calibration and verification of these micro-CMMs. To satisfy these demands, a number of research activities are being carried out at PTB. An overview of some recent developments concerning instrumentation and calibration standards is given in this paper.

An ultra-precision nanopositioning and nanomeasuring machine (NMM) has been upgraded focusing on its mirror corner, interferometers and angle sensors, as well as its weight compensation, its electronic controller, its vibration damping stage and its instrument chamber. The upgrade has significantly reduced its positioning noise, e.g., from $1\sigma = 0.52$ nm to $1\sigma = 0.13$ nm for the z-axis.

Further developments of several microprobes have been detailed, including the boss-membrane piezoresistive probe, the tactile-optical fibre probe and the AFM based probes. The 3D fibre probe has been significantly improved concerning its 3D measurement capability, isotropic probing stiffness and the design of dual-sphere probing styli. The development of a 3D-AFM and of assembled cantilever probes (ACPs) offers full 3D measurements of parts with sizes from a few micrometres to tens of nanometres. This is promising for filling the metrology gap between AFMs and micro-CMMs.

Further developments of calibration standards for force, geometry, contour and microgears have been introduced. A reference spring artefact applicable for calibrating the 3D probing stiffness as well as for characterizing the probing adhesive force has been presented. An Aztec artefact, which applies micromachined micro-pyramidal marks for defining reference coordinates in 3D space, has also been presented. Compared to conventional calibration artefacts, it has advantages such as high surface quality, well-defined geometry and cost-effective manufacturing. A task-specific micro-contour calibration standard for ensuring the traceability, especially of high-precision optical measurements at microgeometries, has been introduced. A workpiece-like microgear standard embodying different gear geometries (modules ranging from 0.1 mm to 1 mm) has been presented.

Unfortunately, despite the large development and research efforts, the take-up of the micro-coordinate measurement technique is currently quite low. The reasons for this are multifold, for instance, the high purchasing and running costs, low measurement throughput, and the lack of generally accepted physical and documentary standards. Therefore, more research and development work is expected in the future with a possible emphasis on more cost-effective instrumentation, better dynamics properties and measurement strategies for higher measurement throughput, multi-sensor and data fusion techniques to merge the advantages of different sensing techniques, as well as the further development of physical and documentary standards.

Acknowledgments: The authors would like to thank Matthias Hemmleb of the m2c company and Denis Dontsov of the SIOS company for their kind help and excellent support.

Author Contributions: G.D. contributed the research results of § 2.1, § 2.4 and § 3.1; S.B. and G.D. contributed that of § 2.2 and § 3.2; U.N.-R., M.N. and M.S. contributed that of § 2.3, § 3.3 and § 3.4, respectively; G.D. wrote the paper with the supports from all co-authors.

Conflicts of Interest: The authors declare no conflict of interest.

References

1. Hansen, H.N.; Carneiro, K.; Haitjema, H.; De Chiffre, L. Dimensional micro and nanometrology. *CIRP Ann. Manuf. Technol.* **2006**, *55*, 721–743.
2. Mattsson, L. Metrology of micro-components—A real challenge for the future. In Proceedings of the 5th International Seminar on Intelligent Computation in Manufacturing Engineering (CIRP ISME '06), Ischia, Italy, 25–28 July 2006; pp. 547–552.
3. Multi-Sensor Metrology for Microparts in Innovative Industrial Products. EMRP Project. Available online: https://www.ptb.de/emrp/microparts-project.html (accessed on 8 April 2016).
4. Peggs, G.N.; Lewis, A.J.; Oldfield, S. Design for a compact high-accuracy CMM. *CIRP Ann. Manuf. Technol.* **1999**, *48*, 417–420.
5. Vermeulen, M.M.P.A.; Rosielle, P.C.J.N.; Schellekens, P.H.J. Design of a high-precision 3D-coordinate measuring machine. *CIRP Ann. Manuf. Technol.* **1998**, *47*, 447–450.
6. Van Riel, M.; Moers, T. Nanometer uncertainty for a micro price. *Mikroniek* **2010**, *50*, 13–17.
7. Jäger, G.; Manske, E.; Hausotte, T.; Büchner, H.J. Laserinterferometrische Nanomessmaschinen. In *Sensoren und Messsysteme 2000. VDI Berichte 1530*; VDI Verlag: Düsseldorf, Germany, 2000; pp. 271–278.
8. Ruijl, T.A.M. Ultra Precision Coordinate Measuring Machine—Design, Calibration and Error Compensation. Ph.D. Thesis, Technische Universität Darmstadt, Darmstadt, Germany, 1 May 2001.
9. ISARA 400 ULTRAPRÄZISIONS-KOORDINATENMESSMASCHINE, Precision Engineering BV. Available online: http://www.ibspe.de/category/isara-400-ultraprazisions-koordinatenmessmaschine.htm (accessed on 8 April 2016).
10. Cao, S.; Brand, U.; Kleine-Besten, T.; Hoffmann, W.; Schwenke, H.; Bütefisch, S.; Büttgenbach, S. Recent developments in dimensional metrology for microsystem components. *Microsyst. Technol.* **2002**, *8*, 3–6.
11. Küng, A.; Meli, F.; Thalmann, R. Ultraprecision micro-CMM using a low force 3D touch probe. *Meas. Sci. Technol.* **2007**, *18*, 319–327.
12. Werth VideoCheck® UA, Werth Messtechnik GmbH. Available online: http://www.werth.de/de/unser-angebot/produkte-nach-kategorie/koordinatenmessgeraete/fuer-labor-und-messraum/werth-videocheck-ua.html (accessed on 8 April 2016).

13. UMAP Vision System Type2 Ultra UMAP404. Mitutoyo Corporation. Available online: http://ecatalog.mitutoyo.com/UMAP-Vision-System-TYPE2-Series-364-Micro-Form-Measuring-System-C1577.aspx (accessed on 8 April 2016).

14. Fan, K.C.; Fei, Y.T.; Yu, X.F.; Chen, Y.J.; Wang, W.L.; Chen, F.; Liu, Y.S. Development of a low cost micro-CMM for 3D micro/nano measurements. *Meas. Sci. Technol.* **2006**, *17*, 524–532.

15. Pril, W.O. Development of High Precision Mechanical Probes for Coordinate Measuring Machines. Ph.D. Thesis, Eindhoven University of Technology, Eindhoven, The Netherlands, December 2002.

16. Bütefisch, S.; Dauer, S.; Büttgenbach, S. Silicon three-axial tactile sensor for the investigation of micromechanical structures. In Proceedings of the 9th International Trade Fair and Conference for Sensors Transducers and Systems, Nürnberg, Germany, 18–20 May 1999; Volume 2, pp. 321–326.

17. Li, R.J.; Fan, K.C.; Miao J, W.; Huang, Q.X.; Tao, S. An analogue contact probe using a compact 3D optical sensor for micro/nano coordinate measuring machines. *Meas. Sci. Technol.* **2014**, *25*, 1–33.

18. Chu, C.L.; Chiu, C.Y. Development of a low-cost nanoscale touch trigger probe based on two commercial DVD pick-up heads. *Meas. Sci. Technol.* **2007**, *18*, 1831–1842.

19. Alblalaihid, K.; Kinnell, P.; Lawes, S.; Desgaches, D.; Leach, R. Performance Assessment of a New Variable Stiffness Probing System for Micro-CMMs. *Sensors* **2016**, *16*, 492.

20. Bos, E.J.C. Aspects of tactile probing on the micro scale. *Precis. Eng.* **2011**, *35*, 228–240.

21. Bauza, M.B.; Hocken, R.J.; Smith, S.T.; Woody, S.C. Development of a virtual probe tip with an application to high aspect ratio microscale features. *Rev. Sci. Instrum.* **2005**, *76*, 095112.

22. Claverley, J.D.; Leach, R.K. Development of a three-dimensional vibrating tactile probe for miniature CMMs. *Precis. Eng.* **2013**, *37*, 491–499.

23. Murakami, H.; Katsuki, A.; Sajima, T.; Suematsu, T. Study of a vibrating fibre probing system for 3-D micro-structures: Performance improvement. *Meas. Sci. Technol.* **2014**, *25*.

24. Guijun, J.; Schwenke, H.; Trapet, E. Opto-tactile sensor. *Qual. Eng.* **1998**, *7–8*, 40–43. (In German)

25. Neuschaefer-Rube, U.; Bremer, H.; Hopp, B.; Christoph, R. Recent developments of the 3D fibre probe. In Proceedings of the 11th Laser Metrology for Precision Measurement and Inspection in Industry, Tsukuba, Japan, 2–5 September 2014.

26. Cui, J.; Chen, Y.; Tan, J. Improvement of dimensional measurement accuracy of microstructures with high aspect ratio with a spherical coupling fibre probe. *Meas. Sci Technol.* **2014**, *25*.

27. Ji, H.; Hsu, H.Y.; Kong, L.X.; Wedding, A.B. Development of a contact probe incorporating a Bragg grating strain sensor for nano coordinate measuring machines. *Meas. Sci. Technol.* **2009**, *20*.

28. Oiwa, T.; Nishitani, H. Three-dimensional touch probe using three fibre optical displacement sensors. *Meas. Sci. Technol.* **2004**, *15*, 80–90.

29. Cui, J.; Li, J.; Feng, K.; Tan, J. Three-dimensional fibre probe based on orthogonal micro focal-length collimation for the measurement micro parts. *Opt Express* **2015**, *23*, 26386–26398.

30. Weckenmann, A.; Hoffmann, J.; Schuler, A. Development of a tunnelling current sensor for a long-range nano-positioning device. *Meas. Sci. Technol.* **2008**, *19*.

31. Michihata, M.; Takaya, Y.; Hayashi, T. Nano position sensing based on laser trapping technique for flat surfaces. *Meas. Sci. Technol.* **2008**, *19*.

32. Claverley, J.D.; Leach, R.K. A review of the existing performance verification infrastructure for micro-CMMs. *Precision Engineering* **2015**, *39*, 1–15.

33. Verein Deutscher Ingenieure. *VDI/VDE 2617 Blatt 12.1 Genauigkeit von Koordinatenmessgeräten—Kenngrößen und Deren Prüfung—Annahme—und Bestätigungsprüfungen* für Koordinatenmessgeräte zum Taktilen Messen von Mikrogeometrien; Beuth Verlag GmbH: Berlin, Germany, 2011. (In German)

34. International Organization for Standardization. *ISO 10360-2:2009-Geometrical Product Specifications (GPS)—Acceptance and Reverification Tests for Coordinate Measuring Machines (CMM)—Part 2: CMMs Used for Measuring Linear Dimensions*; International Organization for Standardization: Geneva, Switzerland, 2009.

35. Thalmann, R.; Meli, F.; Küng, A. State of the art of tactile micro coordinate metrology. *Appl. Sci.* **2016**, *6*.

36. Dai, G.; Bütefisch, S.; Pohlenz, F.; Danzebrink, H.-U. A high precision micro/nano CMM using piezoresistive tactile probes. *Meas. Sci. Technol.* **2009**, *20*.

37. Buetefisch, S.; Dai, G.; Danzebrink, H.-U.; Koenders, L.; Solzbacher, F.; Orthner, M.P. Novel design for an ultra high precision 3D micro probe for CMM applications. In Proceedings of the Eurosensors XXIV Conference, Linz, Austria, 5–8 September 2010.

38. Christoph, R.; Neumann, H.J. *Multisensor Coordinate Metrology*; Verlag Moderne Industrie: München, Germany, 2007; ISBN: 978-3-937889-66-5.

39. Dai, G.; Haeßler-Grohne, W.; Hueser, D.; Wolff, H.; Fluegge, J.; Bosse, H. New developments at Physikalisch-Technische Bundesanstalt in three-dimensional atomic force microscopy with tapping and torsion atomic force microscopy mode and vector approach probing strategy. *J. Micro/Nanolithogr. MEMS MOEMS* **2012**, *11*.

40. Dai, G.; Wolff, H.; Danzebrink, H.U. Atomic force microscope cantilever based microcoordinate measuring probe for true three-dimensional measurements of microstructures. *Appl. Phys. Lett.* **2007**, *91*.

41. Critical Dimension Re-Entrant Tip, Team Nanotec. Available online: http://www.team-nanotec.de/index.cfm?contentid=10&shopAction=showProductDetails&id=399 (accessed on 9 April 2016).

42. Frühauf, J.; Gärtner, E.; Brand, U.; Doering, L. Silicon springs for the calibration of the force of hardness testing instruments and tactile profilometers. In Proceedings of the 4th EUSPEN International Conference, Glasgow, UK, 31 May–2 June 2004; pp. 362–363.

43. Küng, A.; Meli, F. Comparison of three independent calibration methods applied to an ultra-precision micro-CMM. In Proceedings of the 7th International Conference—European Society for Precision Engineering and Nanotechnology, Bremen, Germany, 20–24 May 2007; Volume 1, pp. 230–233.

44. Chao, Z.X.; Tan, S.L.; Xu, G. Evaluation of the volumetric length measurement error of a micro-CMM using a mini sphere beam. In Proceedings of the 9th International Conference on Laser Metrology, Machine Tool, CMM and Robotic Performance, Uxbridge, MA, USA, 30 June–2 July 2009; pp. 48–56.

45. Neuschaefer-Rube, U.; Neugebauer, M.; Ehrig, W.; Bartscher, M.; Hilpert, U. Tactile and optical microsensors: Test procedures and standards. *Meas. Sci. Technol.* **2008**, *19*.

46. Fleischer, J.; Buchholz, I.; Härtig, F.; Dai, G. Mikroverzahnungsprüfkörper μ-DBA für evolventische Profillinienabweichungen. *Tech. Mess.* **2008**, *75*, 168–177.

47. Wedmann, A.; Kniel, K.; Dunovska, V.; Härtig, F.; Klemm, M. Rückführung von Mikroverzahnungsmessungen. In *VDI-Berichte Band 2236*; VDI Verlag: Düsseldorf, Germany, 2014; pp. 199–212. (In German)

48. Dunovska, V.; Krause, M.; Kniel, K. *Messung von Mikroverzahnung—Entwicklung von Verfahren zur Eignungsprüfung von Messgeräten für die Mikroverzahnungsmessung, Forschungsvereinigung Antriebstechnik e.V. (FVA)—Arbeitskreis Messtechnik, Abschlussbericht zu Forschungsvorhaben Nr. 567 II, Heft 1144*; Forschungsvereinigung Antriebstechnik e V. (FVA): Frankfurt, Germany, 2015. (In German)

State of the Art of Tactile Micro Coordinate Metrology

Rudolf Thalmann, Felix Meli and Alain Küng

Abstract: Micro parts are increasingly found in a number of industrial products. They often have complex geometrical features in the millimeter to micrometer range which are not accessible or difficult to measure by conventional coordinate measuring machines or by optical microscopy techniques. In the last years, several concepts of tactile micro coordinate measuring machines have been developed in research laboratories and were partly commercialized by industry. The major challenges were related to the development of innovative micro probes, to the requirements for traceability and to the performance assessment at reduced measurement uncertainty. This paper presents a review on state of the art developments of micro coordinate measuring machines and 3D micro probes in the last 20 years, as far as these were qualified in a comparable way, with a special emphasis on research conducted by the Federal Institute of Metrology METAS in this field. It outlines the accuracy limitations for the probe head including the probing element and for the geometrical errors of the machine axes. Finally, the achieved performances are summarized and the challenges for further research are addressed.

Reprinted from *Appl. Sci.* Cite as: Thalmann, R.; Meli, F.; Küng, A. State of the Art of Tactile Micro Coordinate Metrology. *Appl. Sci.* **2016**, *6*, 150.

1. Introduction

Coordinate measuring machines (CMMs) have become versatile and widespread metrology tools to perform complex dimensional measurement tasks in an efficient way. There is an ongoing trend to miniaturization in mechanical and optical production technology, leading to industrial products with generally larger functionality in a smaller volume and with less energy consumption. Small components with often complex geometrical features are found, e.g., in medical products such as ear implants or hearing aids, in gears of micro motors, in small freeform lenses of mobile phones, in injection systems for the automotive industry, or in the telecom sector for fiber optic or next generation radio frequency technology components. Hence, there is a new demand for highly accurate dimensional measurements on micro parts having geometrical features in the meso scale, *i.e.*, in the sub-millimeter to micrometer range. Such micro parts are often too complex and too large for optical microscopy techniques and their tiny structures are hardly accessible by means of conventional tactile coordinate measuring machines.

The major limiting factors of conventional CMMs for the measurement of small object features are the size of the probing element, the contact force as well as the CMM stage accuracy. Whereas modern multi-sensor CMMs with optical probes, mostly based on imaging systems, are well suited for the measurement of small features, only tactile probes have the fundamental properties needed for high precision full 3D capability with well established traceability. The paper discusses the key elements of tactile micro coordinate metrology, *i.e.*, the probe head for probing objects with small styli and weak forces, the precision stages with an Abbe error free metrology system, as well as the system calibration and performance verification.

Several review papers have been published in the field of micro coordinate metrology, in particular on probing systems [1], on aspects of tactile probing [2] and more recently on existing performance verification infrastructure of micro CMMs [3]. The purpose of this paper is to present a review on state of the art machine and probe developments seeking ultimate accuracy for three-dimensional measurements of micro parts, with a special emphasis on those instruments, where comparable qualification and performance assessment results are available. The research conducted on the micro CMM at METAS [4] is presented in more detail, since this machine has been qualified extensively using methods relying as closely as possible on existing written standards.

2. Probe Developments

When measuring small features using a tactile probe, the influence of the probe-surface interaction gets more and more critical with decreasing size of the probing element. For smaller probe sphere radii, the contact area gets smaller and therefore effects of elastic or even plastic deformation of the surface due to static or dynamic measurement forces become very critical. The effects of surface deformation by over-travel forces due to the probe stiffness and of impact forces due to the probe mass, dynamic excitations due to vibrations, surface forces as well as tip rotations due to stylus bending or rotation has been extensively studied by Bos [2]. Conventional probe systems of CMMs did not fulfil the stringent requirements regarding low stiffness and low dynamic mass nor the accuracy requirements set by ultra precision CMMs, therefore new systems had to be developed. A review over a wide range of probing systems has been given by Weckenmann *et al.* [1].

One the first micro probes has been developed jointly by PTB and Werth in 1998 [5] and soon after implemented on commercial multi-sensor CMMs. It is based on a glass fiber with a quasi spherical melted end whose position is measured with an opto-electronic system. Tip diameters down to 20 μm can be realized. Whereas the first probes had only 2D capability, Werth has developed in the mean time a fiber probe with full 3D capability [6]. The probe is available on Werth multi-sensor CMMs and has proven its usefulness in many applications, in particular for the measurement of small holes. Probing errors of a few hundred nanometers are specified, limited mainly by the relatively large form deviations of the melted glass sphere tips.

PTB developed a probe based on a micro-fabricated silicon membrane with piezo-resistive sensors [7], first published in 1999. The system measures two angular and one translational movement of the probing element. The stiffness in horizontal direction depends on the shaft length and is usually not the same as in the vertical direction. The probe design has been commercialized in the meantime and is available on the F25 micro CMM from Zeiss. The manufacturer specifies a probing error of 250 nm [8].

Approximately at the same time, NPL has also developed a micro-fabricated probe based on three angled flexible elements with capacitive sensors [9]. This original design has been commercialized as the Triskelion probe by IBS [10]. The manufacturer indicates the 3D measurement uncertainty of tip deflection to be below 20 nm. More recently NPL developed a novel vibrating micro-probe, based on the Triskelion design with six piezoelectric sensors and actuators, two on each flexure [11]. The design is optimized to feature high isotropy and very low contact forces in the µN range, thus reducing effects of surface interaction forces [2]. Experimental results to qualify the probe are not yet published.

A similar probe design using three angled flexible elements, but based on a monolithic silicon chip with piezo-resistive sensors and suitable for mass production, has been developed by TUE [12] and first published in 2001. Meanwhile, this is commercialized as the Gannen probe by XPRESS [2,13]. It is specified with a combined 3D probe uncertainty of 45 nm.

Another micro probe which is commercially available was developed by the CMM manufacturer Mitutoyo. The UMAP probe, implemented on the UMAP vision system since 2002 [14], is based on an ultrasonic micro-vibration sensor and available with stylus tip diameters between 15 µm and 300 µm. The accuracy of the probe itself is not specified.

In order to overcome the trade-off between the stiffness requirements of the probing system at contact and during approach, the University of Nottingham is actually developing a probing system for micro CMMs with variable stiffness [15]. While still under development, first promising experiments resulted in an expanded uncertainty of the probing system of about 60 nm. Issues of drift and stylus tip displacement while switching the stiffness yet need to be addressed.

Further probe developments still in the research phase are conducted in Asia: The Hefei University of Technology has been working on several concepts, all based on glass or tungsten styli attached to a floating plate, the latter being suspended on wires or leaf springs [16–18]. The deflection is measured with laser focus sensors or with a laser interferometer combined with an autocollimator based optical sensor for the angular motion. Their latest development includes fiber Bragg grating sensors to measure the deflection of the three arc-shaped cantilevers holding the stylus [19]. The University of Kitakyushu is working on a vibrating fiber probing system [20]. It consists of an extremely fine optical fiber with a glass tip diameter of 5 µm, but it is limited to probing in x/y-direction only.

METAS has developed, in collaboration with the Swiss Federal Institute of Technology EPFL and the industrial partner Mecartex, a 3D probe manufactured out of a single piece of aluminum by conventional milling and electro-discharge machining [4,21]. Its filigree parallel kinematic structure (Figure 1) with 60 µm thick flexure hinges minimizes the moving mass and leaves the probing sphere exactly three translational degrees of freedom. It has a stiffness of 20 N/m and exhibits perfectly isotropic probing forces below 0.5 mN at 20 µm deflection. The probe head allows for exchangeable styli with probe sphere diameters ranging from 0.1 mm to 1 mm. The probe deflection, *i.e.*, the $x/y/z$ components of the translational motion, is measured by three inductive sensors. The tilted orientation of the probe head coordinate system (Figure 1) greatly facilitates the access to the probe head for exchangeable styli and also implicates, that all three axes are affected by the gravitational force in an identical way, allowing for isotropic probing forces even when changing the stylus weight.

Figure 1. METAS tactile probe for micro parts (housing removed).

3. Stage and Metrology Systems Design

The second critical element of an accurate micro CMM is the 3D translation stage with the metrology system. In order to achieve smallest uncertainties, the Abbe principle must be strictly fulfilled, *i.e.*, the reference scale must be in line with the measured length and thus always pointing towards the probe element for all positions of the stage, which imposes severe design restrictions. For classical CMMs with large measurement volumes, this cannot be realized due to limited space and disproportional cost.

Two major classes of metrology systems can be distinguished, *i.e.*, laser interferometer and incremental scale based systems. Abbe free stages with interferometers use a moving frame with mirrors onto which usually the workpiece is placed, as it is shown for the two-dimensional case in Figure 2 (left). Plane mirror interferometers are used to measure the displacement of the mirror frame with respect to the metrology frame with the probe. To build an Abbe free measurement stage using incremental scales is somewhat more complicated because the scales need to be connected with both frames and therefore require additional bearings. Figure 2 (right) shows a possible solution in 2D, as proposed by Vermeulen *et al.* [22]. Whereas measurement stages with incremental scales are more robust for industrial environment and potentially cheaper, they have the disadvantage, that the additional bearing between the two frames is part of the measurement loop and thus induces an additional uncertainty contribution.

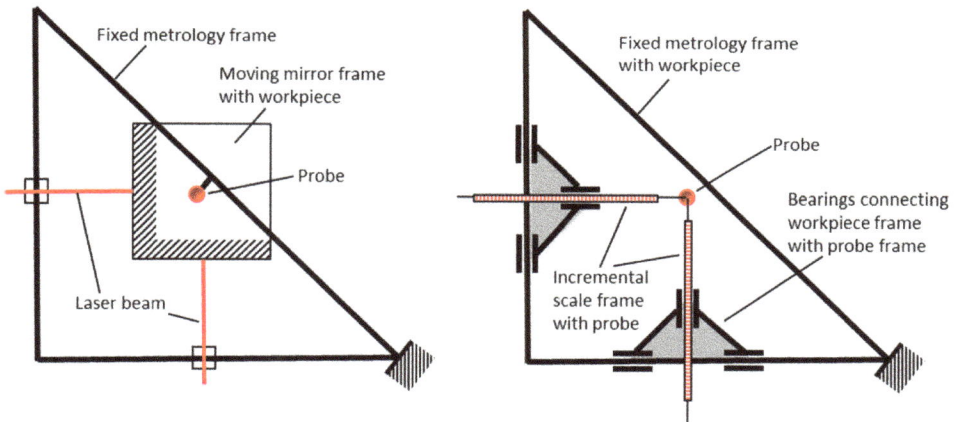

Figure 2. Principle of measurement stages (2D case for simplification of drawing). (**left**) Laser interferometer scales and moving workpiece stage with reference mirror frame; (**right**) Incremental scales and moving probe stage with additional bearings between frames.

Incremental scale based measurement stages have been proposed by TUE [22], which was later on commercialized by Zeiss for the micro coordinate machine F25 [8] (the instrument F25 is no longer part of the product portfolio of Zeiss). In this design, the Abbe principle is respected for the x/y-stage, while for a z-displacement of the probe out of the x/y-plane an Abbe offset must be taken into account. The F25 provides a measurement volume of 100 mm × 100 mm × 100 mm. A similar design, but with all three axes in Abbe, has been realized by TUE [23] for the TriNano micro CMM and is commercialized by XPRESS [13].

One of the first Abbe free interferometric stages for coordinate metrology was realized at NPL: a conventional CMM was used for the 3D movement of a cube corner supporting the mirrors for reflecting the interferometer beams [9]. TU Illmenau designed an ultra-precise 3D stage with Zerodur® (Schott AG, Jena, Germany) frame and a multi-axis interferometer including angular motion (yaw) measurement for their nano-positioning and nano-measuring machine NMM [24], which is commercialized by SIOS [25]. The Hefei University of Technology built two prototype micro CMMs, both with roller bearing stages driven by piezo motors and laser interferometers for stage position sensing [26,27].

An original ultra precision stage development at Philips CFT [28] has been installed at METAS and was the basis for a new development by IBS [29] leading to the Isara 400 machine [30]. The ultraprecision stage of the METAS micro CMM [4,28] has vacuum preloaded air bearings. The stages are driven by Lorenz actuators and their motion is measured and controlled by three plane mirror heterodyne interferometers. The original V-configuration of the guideways with wedges as shown in Figure 3 makes the stage very compact and stiff. The working volume of the stage is 90 mm × 90 mm × 38 mm. According to Figure 2 (left) the probe head remains in a fixed position connected to the metrology frame while the stage is moving the workpiece during measurements. The workpiece is located in a Zerodur cube corner with three perpendicular flat mirrors forming the reference coordinate system. All three interferometer beams measuring the workpiece displacement are pointing to the center of the probing sphere, thus limiting the residual Abbe offset to the probe radius.

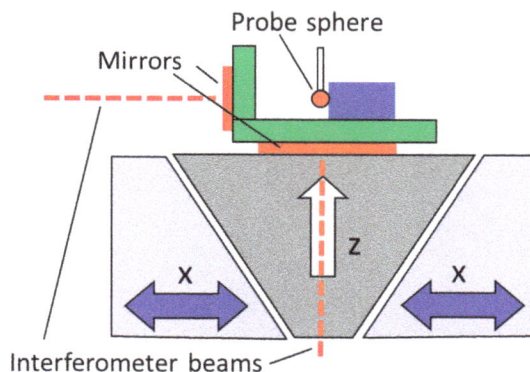

Figure 3. Principle of the Abbe free CMM stage at METAS. y-axis perpendicular to drawing plane.

4. System Calibration and Performance Verification

4.1. Probe Calibration

4.1.1. Single Point Probing

The single point probing behavior of a CMM probe can be characterized by the P_{FTU} value (single stylus form error) as specified in the standard ISO 10360-5 [31], based on the peak-to-valley deviation of 25 probing points on a sphere arranged according to Figure 4 (left). For two of the micro CMMs presented above, corresponding results are available. The P_{FTU} value for the Zeiss F25 probe was found from the average of 14 measurements to be 189 nm with 60 nm standard deviation and all values but one complying with the manufacturer specification of 250 nm [8]. A repeatable pattern was found in these measurements, most probably due to the anisotropic behavior of the silicon chip membrane as suggested by Bergmans *et al.* [8].

METAS has carried out the same standardized test [31] for single point probing error on a 1 mm diameter sphere [32]. The test was performed four times with the sphere rotated by 90° between each test. The sphere radius was found to be between 0.500202 nm and 0.500211 nm with a standard deviation of 4 nm. The test resulted in P_{FTU} values between 53 nm and 67 nm with an average value of 58 nm. The measurements in Figure 4 (right) show a systematic behavior probably due to an uncompensated form deviation of the probing sphere. The Si_3N_4 test sphere [33] had an estimated form deviation of less than 15 nm.

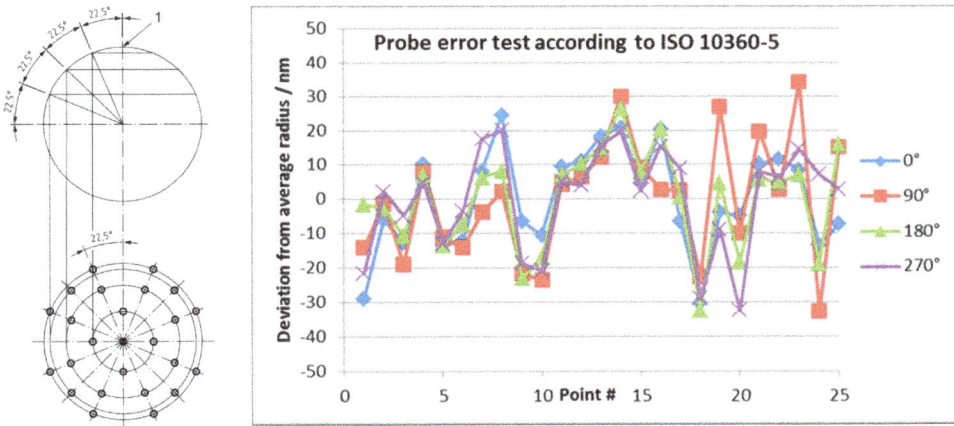

Figure 4. Determination of the single point probing error according to ISO 10360-5 on a 1 mm Si_3N_4 sphere: (**left**) Arrangement of probing points on the reference sphere; (**right**) Deviation from average radius for four measurements with sphere rotated by 90°.

31

4.1.2. Scanning

The standard ISO 10360-4 [34] describes the scanning probing error T_{ij} which is the range of radial distances obtained from four circular scans on a sphere: one full circle in an equatorial plane, a second full circle in a plane parallel to the first, a third half circle through the pole, and a fourth half circle in a plane perpendicular to the third at a distance of about 1/2 of the radius from the pole (Figure 5, left).

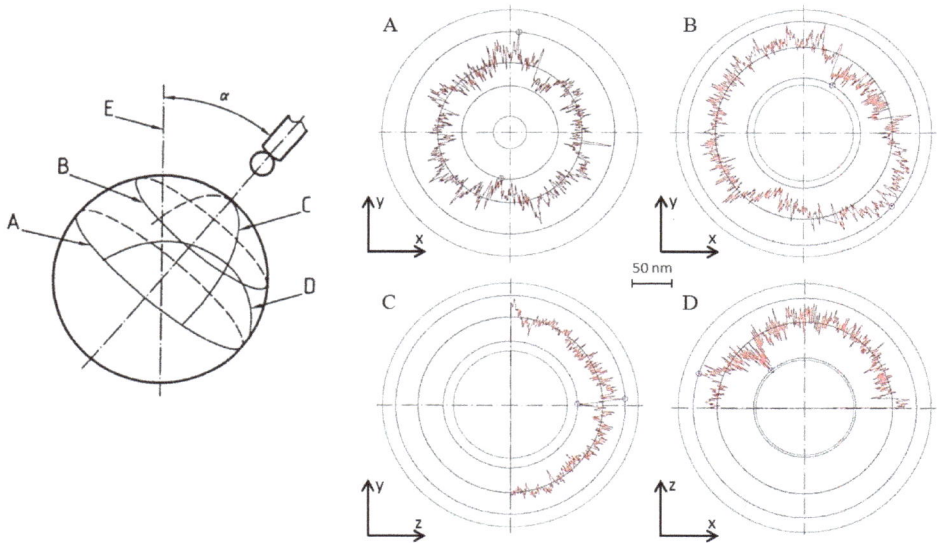

Figure 5. Determination of the single point probing error according to ISO 10360-4 on a 1 mm ruby sphere: (**left**) Arrangement of scanning profile planes A, B, C and D; (**right**) Measured profiles A to D resulting in a scanning probing error T_{ij} = 87 nm.

This test was performed with the METAS probe on a 1 mm ruby sphere with a point density of 300 pts/mm [35]. The LS diameter fitted through the scan profiles resulted in 1.000806 mm, compared to the value of 1.000804 mm found by the independent probe sphere diameter determination as shown in Section 4.1.3. For T_{ij} a value of 87 nm was obtained. The corresponding values for the four individual scans were 51 nm, 75 nm, 67 nm and 78 nm, respectively. It must be noted that these values comprise the form deviation of the reference sphere (53 nm), the uncompensated form deviation of the probing sphere and surface roughness contributions of both spheres, as no filtering was applied.

4.1.3. Probe Sphere Diameter and Form Correction

On a conventional CMM the absolute diameter of the probe sphere is calibrated on a reference sphere. At the same time the anisotropic behavior of the probe head including stylus bending is corrected. This requires an independent calibration of the diameter of the reference sphere and assumes the roundness deviation of both, the reference sphere and the probe sphere, to be much smaller than the overall probing error, which is generally the case. For high precision micro coordinate metrology, the requirements for the reference sphere are much higher and cannot be fulfilled by independent calibrations. Therefore, an error separation method was developed involving three nominally identical spheres compared against each other in various configurations, using one of the spheres both as the probe sphere and the reference sphere, respectively [36]. In this way, both the absolute diameter and the sphericity deviation of three 1 mm ruby spheres could be mapped over their entire accessible surface. This method finally relies directly on the traceability of the CMM stage interferometer and on the repeatability of the probe being usually below 4 nm (standard deviation). Figure 6 shows a sphericity map of one of the measured spheres with a form deviation of 33 nm.

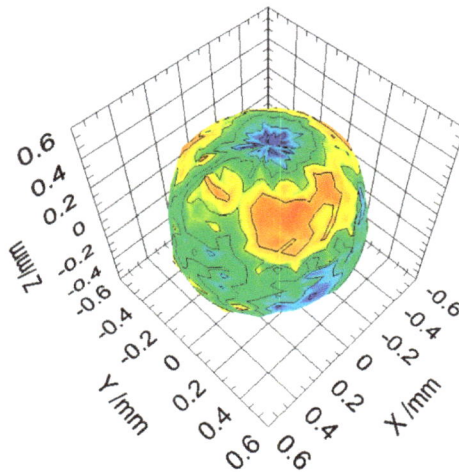

Figure 6. Sphericity map of a 1 mm ruby sphere with 33 nm form deviation. The 3D plot shows results obtained by probing the surface with a 10° resolution in longitude and a 5° resolution in latitude.

4.2. Stage Calibration and Error Correction

The way of calibrating the measurement stage depends much on its design (Section 3). Measurement stages for micro CMMs with incremental scales are calibrated and corrected similarly to conventional machines, *i.e.*, the position accuracy is measured with a laser interferometer and the geometrical errors are usually determined with the help of artefacts. For the F25 machine some measurement results were published by PTB [37]. They carried out straightness measurements on a Zerodur gauge block assembly along the horizontal y-axis over 43 mm, resulting in a straightness deviation of about 30 nm with a reproducibility that would allow for a further correction down to a few nanometers residual error. A similar measurement along the vertical z-axis over a length of 1.75 mm (limited by the length of the stylus shaft) resulted in a straightness deviation of less than 10 nm.

The calibration of interferometer based stages includes the calibration of the laser wavelength (given by the optical frequency and the refractive index determined by the compensation unit), and—more critically—the flatness and squareness of the faces of the reference mirror cube. For the flatness calibration of the reference mirrors, high precision reference artefacts are used. IBS reported on flatness measurements using a Zerodur block with a metallic coating allowing for a capacitive sensor to be used instead of the tactile probe [29] und thus improving resolution and noise due to surface roughness and probe repeatability. The squareness calibration was made using the same artefact by applying a reversal technique.

The calibration and correction of the Zerodur reference mirror of the METAS micro CMM has been made in a similar way as described above [35]. For this, a Zerodur block of 100 mm × 60 mm × 40 mm with form deviations <10 nm was used. The gold coated block was scanned by a capacitance sensor. Where possible, reversal techniques were applied to separate the errors. The resulting errors were then mapped in correction files. The calibration of the squareness in x/y-plane was made with the help of a ball plate. The squareness of the z-axis with respect to horizontal axes was calibrated with the help of a Ø 44 mm tungsten carbide sphere, by scanning the sphere in two half-circles in the x/z- and the y/z-planes and adjusting the corresponding orthogonality angles as to obtain minimum residuals from a circle. Although somewhat less sensitive than checking the orthogonality by, e.g., measuring a ball bar in diagonal directions, the method using a sphere is more efficient and provides an independent performance check.

4.3. Performance Tests

Performance tests of micro CMMs can be made in the same way as for conventional machines following the procedure of the acceptance and reverification test ISO 10360-2 [38] using specially designed small artefacts. Since the target uncertainties are in the nanometer range, the stability of these artefacts is a particular challenge. The quantity to be determined in the above cited ISO test is E_0, the error of indication of a calibrated test length.

Ball bars of different lengths ranging from 20 mm to 100 mm were manufactured using Zerodur bars and ruby spheres (Figure 7). The results shown in Figure 7 assess the E_0 value of the machine in the whole volume to be 27 nm + $0.2 \times 10^{-6} \times L$ [35]. Since the test requires the ball bar to be measured along the diagonals of the volume, one can compute the squareness between the three machine axes. It is an accurate method since the longest diagonal length can be measured.

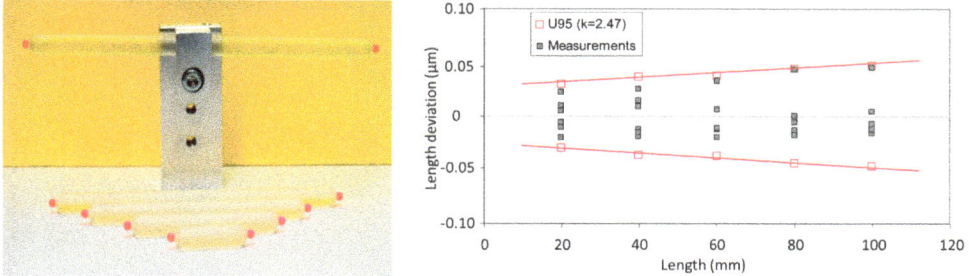

Figure 7. (**left**) Micro ball bars used for performance verification; (**right**) Length deviations measured with 5 different ball bars each oriented in different axial and diagonal directions. Red squares denote the symmetric 95% confidence intervals, red lines the fitted 95% confidence length measurement error E_0 = 27 nm + $0.2 \times 10^{-6} \times L$.

Ball plates are another suitable artefact to test the performance of a CMM. The calibration method requires measurements in 4 positions (two from each side of the plate) in order to be able to separate the errors of the CMM from the deviations of the ball plate nominal positions. The ball plate used in the test hereafter was manufactured using 25 ruby spheres, Ø 3 mm, pressed into the holes of an 85 mm × 85 mm invar frame (Figure 8).

Apart from the squareness error between the axes, as it is also obtained from ball bar measurements described above, the method also provides information about other error contributions, such as axes straightness or angular distortions. As shown in Figure 8 the measurement results on the METAS micro CMM exhibit a residual squareness error of (0.17 ± 0.05)". The contributions from other error sources, which are smaller than 40 nm, are largely dominated by temperature drifts. Evaluating the 300 relative distances between all measured ball center positions in a similar way as in Figure 7 (right) leads to a 95% confidence length measurement error of E_0 = 26 nm + $0.23 \times 10^{-6} \times L$, a value very close to the one obtained using the ball bars.

PTB and METAS carried out a bilateral comparison on a 90 mm × 90 mm Zerodur plate with hemispheres [39]. The average difference of the hemisphere x- and y-coordinates was 12 nm and 7 nm respectively, with a maximum value of 28 nm. It must be noted that, since an error separation calibration procedure was applied, these measurements do not represent the deviations of the machine, but only the capability and comparability of calibrating a ball plate standard.

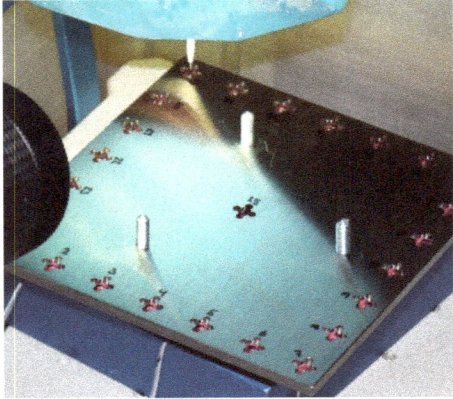

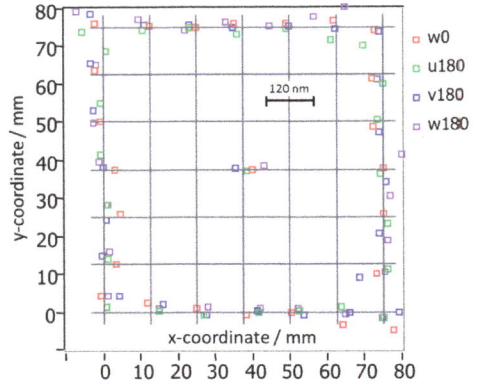

Figure 8. (left) Micro ball plate used for performance verification; **(right)** Deviations measured for 25 ball center positions in four different orientations of the ball plates. The evaluation of all relative ball distances results in a 95% confidence length measurement error of $E_0 = 26$ nm $+ 0.23 \times 10^{-6} \times L$.

5. Challenges for Further Development

To extend the use of tactile micro coordinate metrology for a large variety of industrial applications, a series of challenges need to be addressed.

5.1. Accessibility of 3D Features

Quite often, the geometrical features of a micro-parts are not easily accessible with a spherical probe on a straight stylus, but limited by the probe diameter for small holes, by the stylus length for high aspect features such as deep holes, or by the direction of the probe stylus. Therefore, smaller probe elements with smaller and better known form deviation, longer styli to access deep holes and smaller probing forces to allow for longer styli need to be developed. To achieve a better accessibility of normally hidden features, probes with multiple, *i.e.*, star-shaped styli were built [40], their application turned out to be delicate for standard use, however. Another approach is to implement additional rotational axes on the micro CMM [41], obviously at the cost of accuracy due to uncorrected guide and positional errors of the rotation axes.

5.2. Probe/Surface Interaction

The smaller the workpiece features, the probe sphere diameter and the targeted measurement uncertainty are, the more important becomes the interaction of the probe with the surface. Not only effects of surface indentation due to measurement forces [2], but also stylus wear and deposition of wear debris on the probe sphere become relevant [42]. These effects can be significantly reduced by selection of appropriate materials for the probe sphere, *e.g.*, diamond coated spheres [43].

For very weak probing forces, effects of attractive forces between stylus tip and surface get relevant and may cause sticking effects [2,11]. These problems may be overcome by novel concepts such as vibrating probes [11,14,20].

5.3. Task Specific Uncertainty

Performance verification tests as described in Section 4 may result in standardized length measurement errors and qualify the probe errors, but do provide only some basic contributions to the uncertainty of a specific geometrical feature to be determined. Length measurement errors are certainly not adequate for the uncertainty of derived geometrical elements, such as a local radius, a form deviation or an angle. Methods for the estimation of reliable, task specific measurement uncertainties still need to be developed further and are not yet mature for day to day applications. A promising approach is the *Virtual CMM* based on numerical simulation [44,45], which has been successfully implemented on the METAS micro CMM [46], but which still requires a considerable effort in modelling and computing. Further research is needed for refining and implementing the error model adapted to the particular characteristics of a micro CMM and integrating this directly into a high level software, allowing to achieve estimations of uncertainty not only off-line, but with much higher computing speed almost in real-time during the measurement process.

5.4. Application in Industry

For a wider use of tactile micro coordinate metrology in industry, several issues need to be further addressed. The need for fast and efficient measurements contradicts to the requirements of slow approaching speed to achieve small contact forces and thus asks for improvements of the dynamic properties of machines and probes. Another problem is the fragility of probing systems and styli, which easily break due to overtravel or uncontrolled manipulation. Systems with the possibility of exchangeable styli obviously help to save costs of replacing the entire probe system. Finally, industry is asking for comparability of measurement results and of system specifications. A written standard for performance verification dedicated to micro CMMs, or preferably an amendment of an existing standard is clearly needed [3].

6. State of the Art and Conclusions

As outlined earlier there are actually a few types of machines for high precision tactile micro coordinate measurements available, which have full 3D measurement capability and are operational with a high-level software platform to provide industrial measurement services: the well established Zeiss F25 [8], from which about 25 machines were sold and are operating in research labs and institutes worldwide, but is unfortunately no longer commercially available; two more recent developments, the Isara 400 [30] and the TriNano [13,23] are both in principle commercially available, but the market uptake did not really happen so far; the micro CMM at METAS [4] is a prototype development, which is fully operational since 2006 and almost daily used for providing calibration and measurement services to industry. Also to be mentioned is the commercially available SIOS nanopositioning and nanomeasuring machine [24,25], which can be configured with a suitable probe to perform 3D coordinate measurements. Furthermore, micro CMMs were developed by Hefei University of Technology [26,27], complemented by tactile micro probe developments [16–19] to achieve 3D measurement capability. These latter laboratory instruments, however, are still in the development phase and have not proven yet their full measurement capability following standardized procedures as outlined in Section 4. Not discussed further in this context are multi-sensor CMMs, such as [6,14], which have accuracies closer to conventional CMMs.

State of the art probes of above mentioned instruments do achieve a repeatability of a few nanometers, but the best values reported for the 3D probing error according to standardized specification tests [31,34] are around 50 nm, with similar performance for single point probing and scanning. The achievable probing errors are limited mainly by the form deviation and the roughness of the probing sphere, and the overall stability of the instrument during the test procedure. State of the art CMM stages were demonstrated to have maximum errors of indicated length according to a standardized specification test [38] in the entire volume and in arbitrary direction of about 60 nm, limited by the uncorrected flatness deviation of the reference mirrors, residual squareness errors and the errors of the probe, the latter being part of the mentioned specification test.

In conclusion, in spite of the big research and development effort spent worldwide for tactile micro coordinate metrology, there are still very few results available for a comparable performance verification according to standardized procedures. The reason for this is partly that many instruments and probe systems are still in the development phase, but there is also a lack of generally accepted and widely available material standards and written standards dedicated to the particular characteristics of micro CMMs.

Acknowledgments: This work was funded by the R & D-Program at METAS and the European Metrology Research Program EMRP.

Author Contributions: M.F., A.K. and R.T. conceived and designed the experiments; A.K. performed the experiments and analyzed the data; R.T. wrote the paper.

Conflicts of Interest: The authors declare no conflict of interest.

Abbreviations

The following abbreviations are used in this manuscript:

CMM   Coordinate measuring machine
3D   Three dimensional

References

1. Weckenmann, A.; Estler, T.; Peggs, G.; McMurtry, D. Probing systems in dimensional metrology. *Ann. CIRP* **2004**, *53*, 657–684.
2. Bos, E.J.C. Aspects of tactile probing on the micro scale. *Precis. Eng.* **2011**, *35*, 228–240.
3. Claverley, J.D.; Leach, R.K. A review of the existing performance verification infrastructure for micro-CMMs. *Precis. Eng.* **2014**, *39*, 1–15.
4. Küng, A.; Meli, F.; Thalmann, R. Ultraprecision micro-CMM using a low force 3D touch probe. *Meas. Sci. Technol.* **2007**, *18*, 319–327.
5. Guilun, J.; Schwenke, H.; Trapet, E. Opto-tactile sensor for measuring small structures on coordinate measuring machines. *Proc. ASPE* **1998**, *18*, 25–28.
6. Werth 3D Fibre Probe. Available online: http://www.werth.de/index.php?id=262&L=1 (accessed on 26 April 2016).
7. Buetefisch, S.; Dai, G.; Danzebrink, H.-U.; Koenders, L.; Solzbacher, F.; Orthner, M.P. Novel design for an ultra high precision 3D micro probe for CMM applications. *Procedia Eng.* **2010**, *5*, 705–712.
8. Bergmans, R.H.; Nieuwenkamp, H.J.; van Veghel, M.G.A. Probing behavior of a microCMM. In Proceedings of the 11th Euspen International Conference, Como, Italy, 23–26 May 2011; Volume 1, pp. 104–107.
9. Peggs, G.N.; Lewis, A.; Oldfield, S. Design of a compact high-accuracy CMM. *Ann. CIRP* **1999**, *48*, 417–420.
10. IBS Precision Engineering. Available online: http://www.ibspe.com/category/isara-400-3d-cmm/triskelion-touch-probe.htm (accessed on 26 April 2016).
11. Claverley, J.K.; Leach, R.K. Development of a three-dimensional vibrating tactile probe for miniature CCMs. *Precis. Eng.* **2013**, *37*, 491–499.
12. Haitjema, H.; Pril, W.O.; Schellekens, P.H.J. Development of a silicon-based nanoprobe system for 3-D measurements. *Ann. CIRP* **2001**, *50*, 365–368.
13. XPRESS Precision Engineering B.V. Available online: http://www.xpresspe.com/probe2.php (accessed on 26 April 2016).
14. Mitutoyo. Available online: http://www.mitutoyo.co.jp/eng/new/news/2002/02_13.html (accessed on 26 April 2016).
15. Alblalaihid, K.; Kinnell, P.; Lawes, S.; Desgaches, D.; Leach, R. Performance Assessment of a New Variable Stiffness Probing System for Micro-CMMs. *Sensors* **2016**, *16*, 492.

16. Fan, K.C.; Cheng, F.; Wang, W.; Chen, Y.; Lin, J.Y. A scanning contact probe for a micro-coordinate measuring machine (CMM). *Meas. Sci. Technol.* **2010**, *21*, 054002.

17. Li, R.J.; Fan, K.C.; Miao, J.W.; Huang, Q.X.; Tao, S.; Gong, E.R. An analogue contact probe using a compact 3D optical sensor for micro/nano coordinate measuring machines. *Meas. Sci. Technol.* **2014**, *25*, 094008.

18. Li, R.J.; Fan, K.C.; Huang, Q.X.; Zhou, H.; Gong, E.R.; Xiang, M. A long-stroke 3D contact scanning probe for micro/nano coordinate measuring machine. *Precis. Eng.* **2016**, *43*, 220–229.

19. Liu, F.F.; Chen, L.J.; Wang, J.F.; Xia, H.J.; Li, R.J.; Yu, L.D.; Fei, Y. Modeling and prototyping of a fiber Bragg grating-based dynamic micro-coordinate measuring machine probe. *Meas. Sci. Technol.* **2016**, *27*, 025016.

20. Murakami, H.; Katsuki, A.; Sajima, T.; Suematsu, T. Study of a vibrating fiber probing system for 3-D micro-structures: Performance improvement. *Meas. Sci. Technol.* **2014**, *25*, 094010.

21. Meli, F.; Fracheboud, M.; Bottinelli, S.; Bieri, M.; Thalmann, R.; Breguet, J.-M.; Clavel, R. High precision, low force 3D touch probe for measurements on small objects. In Proceedings of the Euspen International Topical Conference, Aachen, Germany, May 2003; pp. 411–414.

22. Vermeulen, M.M.P.A.; Rosielle, P.C.J.N.; Schellekens, P.H.J. Design of a high-precision 3D-coordinate measuring machine. *Ann. CIRP* **1998**, *47*, 447–450.

23. Van Riel, M.; Moers, T. Nanometer uncertainty for a micro price. *Mikroniek* **2010**, *50*, 13–17.

24. Jäger, G.; Manske, E.; Hausotte, T.; Buchner, H.-J. The metrological basis and operation of nanopositioning and nanomeasuring machine. *Tech. Mess.* **2009**, *76*, 227–234.

25. SIOS Meßtechnik GmbH. Available online: http://www.sios-de.com/?page_id=635 (accessed on 26 April 2016).

26. Fan, K.C.; Fei, J.T.; Yu, X.F.; Chen, Y.J.; Wang, W.L.; Chen, F.; Liu, S.L. Development of a low-cost micro-CMM for 3D micro/nano measurements. *Meas. Sci. Technol.* **2006**, *17*, 524–532.

27. Huang, Q.X.; Wu, K.; Wang, C.; Li, R.J.; Fan, K.C.; Fei, J.T. Development of an Abbe Error Free Micro Coordinate Measuring Machine. *Appl. Sci.* **2016**, *6*, 97.

28. Ruijl, T. Ultra Precision Coordinate Measuring Machine; Design, Calibration and Error Compensation. Ph.D. Thesis, Technical University of Delft, Delft, The Netherlands, February 2001.

29. Widdershoven, I.; Spaan, H.A.M. Calibration of the ISARA 400 ultra-precision CMM. In Proceedings of the First International Workshop on Advances in IT—Service Process Engineering, Guadeloupe, France, 23–28 February 2011.

30. IBS Precision Engineering. Available online: http://www.ibspe.com/category/isara-400-3d-cmm.htm (accessed on 26 April 2016).

31. ISO 10360-5. *Geometrical Product Specifications (Gps)—Acceptance and Reverification Tests for Coordinate Measuring Machines (Cmm)—Part 5: Cmms Using Single and Multiple Stylus Contacting Probing Systems*; International Organization for Standardization: Geneva, Switzerland, 2010.

32. Küng, A. *Towards Truly 3D Metrology for Advanced Micro-Parts.* EURAMET Project 1088. 2012. Available online: http://www.euramet.org/technical-committees/length/tc-l-projects/ (accessed on 26 April 2016).

33. Saphirwerk. Available online: http://www.saphirwerk.com/ (accessed on 26 April 2016).

34. ISO 10360-4. *Geometrical Product Specifications (Gps)—Acceptance and Reverification Tests for Coordinate Measuring Machines (Cmm)—Part 4: Cmms Used in Scanning Measurement Mode*; International Organization for Standardization: Geneva, Switzerland, 2000.

35. Küng, A.; Meli, F. Comparison of three independent calibration methods applied to an ultra-precision μ-CMM. In Proceedings of the 7th Euspen International Conference, Cranfield, UK, 20–24 May 2007; Volume 1, pp. 230–233.

36. Küng, A.; Meli, F. Self calibration method for 3D roundness of spheres using an ultraprecision coordinate measuring machine. In Proceedings of the 5th Euspen International Conference, Montpellier, France, May 2005; pp. 193–196.

37. Neugebauer, M. Precision size and form measurements with a micro-CMM F25. In Proceedings of the IXth International Scientific Conference Coordinate Measuring Technique, Ustroń, Poland, 14–16 April 2010.

38. ISO 10360-2. *Geometrical Product Specifications (GPS)—Acceptance and Reverification Tests for Coordinate Measuring Machines (CMM)—Part 2: CMMs Used for Measuring Linear Dimensions*; International Organization for Standardization: Geneva, Switzerland, 2009.

39. Neugebauer, M. Bilateral Comparison on Micro-CMM Artefacts between PTB and METAS; EURAMET Project 1105 Final Report 2011. Available online: http://www.euramet.org/get/?tx_stag_base%5Bfile5D=3457&tx_stag_base%5Baction%5D=download Raw&tx_stag_base%5Bcontroller%5D=Base (accessed on 26 April 2016).

40. Küng, A.; Meli, F. Versatile probes for the METAS 3D Micro-CMM. In Proceedings of the 8th Euspen International Conference, Zürich, Switzerland, 18–22 May 2008; Volume 1, pp. 269–272.

41. Küng, A.; Meli, F.; Nicolet, A. A 5 degrees of freedom μCMM. In Proceedings of the 14th Euspen International Conference, Dubrovnik, Croatia, 2–6 June 2014; Volume 1, pp. 269–272.

42. Küng, A.; Nicolet, A.; Meli, F. Study of sapphire probe tip wear when scanning on different materials. *Meas. Sci. Technol.* **2012**, *23*, 094016.

43. Küng, A.; Nicolet, A.; Meli, F. Study of wear of diamond-coated probe tips when scanning on different materials. *Meas. Sci. Technol.* **2015**, *29*, 084005.

44. ISO/TS 15530-4. *Geometrical Product Specifications (GPS)—Coordinate Measuring Machines (CMM) Technique for Determining the Uncertainty of Measurement—Part 4: Evaluating Task-Specific Measurement Uncertainty Using Simulation*; International Organization for Standardization: Geneva, Switzerland, 2008.

41

45. Härtig, F.; Trapet, E.; Wäldele, F.; Wiegand, U. Traceability of coordinate measurements according to the virtual CMM concept. In Proceedings of the 5th IMEKO TC-14 Symposium on Dimensional Metrology in Production and Quality Control, Zaragoza, Spain, 25–27 October 1995; pp. 245–254.
46. Küng, A.; Meli, F.; Nicolet, A.; Thalmann, R. Application of a virtual coordinate measuring machine for measurement uncertainty estimation of aspherical lens parameters. *Meas. Sci. Technol.* **2014**, *25*, 094011.

A New Kinematic Model of Portable Articulated Coordinate Measuring Machine

Hui-Ning Zhao, Lian-Dong Yu, Hua-Kun Jia, Wei-Shi Li and Jing-Qi Sun

Abstract: Portable articulated coordinate measuring machine (PACMM) is a kind of high accuracy coordinate measurement instrument and it has been widely applied in manufacturing and assembly. A number of key problems should be taken into consideration to achieve the required accuracy, such as structural design, mathematical measurement model and calibration method. Although the classical kinematic model of PACMM is the Denavit-Hartenberg (D-H) model, the representation of D-H encounters the badly-conditioned problem when the consecutive joint axes are parallel or nearly parallel. In this paper, a new kinematic model of PACMM based on a generalized geometric error model which eliminates the inadequacies of D-H model has been proposed. Furthermore, the generalized geometric error parameters of PACMM are optimized by the Levenberg-Marquard (L-M) algorithm. The experimental result demonstrates that the measurement of standard deviation of PACMM based on the generalized geometric error model has been reduced from 0.0627 mm to 0.0452 mm with respect to the D-H model.

Reprinted from *Appl. Sci.* Cite as: Zhao, H.-N.; Yu, L.-D.; Jia, H.-K.; Li, W.-S.; Sun, J.-Q. A New Kinematic Model of Portable Articulated Coordinate Measuring Machine. *Appl. Sci.* **2016**, *6*, 181.

1. Introduction

The portable articulated coordinate measuring machine (PACMM) is a kind of high accuracy coordinate measurement instrument and has the advantages of flexibility, lightweight, portability and easy use compared to the traditional orthogonal CMM. It has been widely applied in manufacturing, assembly [1], *in-situ* measurement, reverse engineering and calibration [2,3]. Error sources of PACMM include structural parameters errors, joint errors, link deflections, thermal deformations and so on. The measurement accuracy of PACMM depends on the proper kinematic model which considers both geometric and non-geometric errors.

Many methods have been proposed in the literatures to establish the kinematic model of robot or PACMM. The Denavit-Hartenberg (D-H) [4] formulation is regarded as the most classical kinematic model for a robot or PACMM. A modified four-parameter D-H formulation [5,6] has been proposed to overcome the badly-conditioned problems when the two adjacent joint axes are parallel or nearly parallel. Reference [7] also showed that another type of the modified D-H formulation which the standard D-H conventions post-multiplies the rotation term around Y-axis could improve the badly-conditioned problem when the two adjacent joint axes are parallel or nearly parallel. In order to meet the three principles of the kinematic model of the manipulator—parametric, continuity and completeness, references [8,9] proposed a complete, parametrically and continuous kinematic model by adding two parameters to Roberts' line parameters. Recently, some researchers also proposed the generalized geometric error model method considering both geometric and non-geometric errors. For example, reference [10] introduced the generalized geometric error parameters for eliminating the geometric and non-geometric errors and improving the positioning accuracy of the patient positioning system. A general approach for error model of machine tools [11] has been introduced to eliminate the geometric and non-geometric errors of machine tools.

The calibration technique of kinematic structural parameters has also been considered as an efficient method of eliminating geometric and non-geometric errors of CMM, PACMM, machine tools, *etc.* According to performance tests of PACMM [12–14], it is required to calibrate the measurement volume of PACMM by using a 1D standard gauge. Reference [15] introduced a simple artifact with two spheres for the kinematic structural parameters calibration of PACMM. The center distance between two spheres in the artifact was calculated by fitting the central points of two spheres. Reference [16] reported 1D ball bar array. The central points of two spheres were measured by PACMM with the special rigid probe. Reference [17] presented a new full pose measurement method for the kinematic structural parameters calibration of the serial robot. This approach was achieved by an analysis of the features of a set of target points on circular trajectory. Kovač *et al.* [18] developed a high accuracy measurement device based on laser interferometer combining with 1D translation table for calibration and verification of PACMM. Shimojima *et al.* [19] presented a 3D ball plate with nine balls. Piratelli *et al.* [20] introduced the development of virtual ball bar to evaluate the performance of PACMM. González *et al.* [21] introduced a virtual circle gauge was applied in evaluating the performance of PACMM and it was composed of bar gauges of 1000 mm length with four groups of three cone-shaped holes. The above-mentioned virtual geometric gauges are applied to reduce the number of test positions, avoid the measurement points randomly sampled on the virtual geometrical gauges surfaces according to the norms and improve the efficiency of verification procedure for PACMM. Acero *et al.* [22] presented an indexed metrology platform combined with a calibrated ball bar gauge, which was applied in evaluating the performance of PACMM. A simplified and low-price length gauge [23] was applied in calibrating the kinematic structural parameters of PACMM and it had three degrees of freedom (DOFs) and a coefficient of low-thermal expansion.

Besides, calibration algorithm of the robot or PACMM plays a crucial role in improving the measurement accuracy of PACMM. D-H conventions of the measuring arm were improved by Genetic Algorithm (GA) [24]. Although GA had good global search ability during the optimization process of D-H parameters of PACMM, it also had poor local search ability. The kinematic structural parameters of parallel dual-joint CMM were calibrated by using Particle Swarm Optimization (PSO) algorithm [25]. Compared with GA, in most cases, all the particles of PSO may converge to the optimal solution more quickly, but it also may be easy to fall into local optimum. Gao *et al.* [26] proposed a modified Simulated Annealing (SA) algorithm for identifying the structural parameters of PACMM. To overcome the disadvantages of the above-mentioned algorithms which have slow converge rate, the nonlinear least square method [27] has been adopted to calibrate the geometric errors of flexible coordinate measuring robot and compensate its geometric errors. The L-M algorithm proposed by Levenberg and Marquardt [28,29] can improve the disadvantages of the nonlinear least-square method by over-relying on the initial values and having high converge rate. Therefore, L-M algorithm is selected as the calibration algorithm of the generalized geometric error parameters of PACMM.

2. Model

The measurement model aims to establish the transformation relationship between the base frame and the center of the probe, and the model includes the nominal geometric parameters, geometric and non-geometric errors. D-H conventions and generalized geometric error theory are applied in establishing the kinematic model of PACMM, respectively.

2.1. D-H Conventions

D-H conventions is supposed to as the most classical kinematic model method for a robot or PACMM. The transformation matrix A_i from the coordinate frame "$i-1$" to "i" with D-H conventions is indicated by Equation (1):

$$A_i = \begin{bmatrix} cos\theta_i & -sin\theta_i cos\alpha_i & sin\theta_i sin\alpha_i & l_i cos\theta_i \\ sin\theta_i & cos\theta_i cos\alpha_i & -cos\theta_i sin\alpha_i & l_i sin\theta_i \\ 0 & sin\alpha_i & cos\alpha_i & d_i \\ 0 & 0 & 0 & 1 \end{bmatrix} \tag{1}$$

where θ_i, l_i, α_i and d_i denote the joint angle, link length, twist angle, joint offset, respectively.

2.2. Generalized Geometric Error Theory

To describe the kinematic model of PACMM, the transformation matrix of the coordinate frame f_i^{ideal} with respect to f_{i-1}^{real} is indicated by using D-H's 4×4 matrix $_i$. Figure 1 shows the transformation relationship between the coordinate frame f_{i-1}^{real} and f_i^{real}. However, the actual geometric parameters of the coordinate frame f_i^{ideal} with respect to f_i^{real} for PACMM exists the slightly deviations from the nominal values because of the existence of machining errors, assembly errors, link deformation and so on. The transformation relationship between the coordinate frame f_i^{ideal} and f_i^{real} is shown in Figure 2. Therefore, the coordinate frame f_{i-1}^{real} transformed to f_i^{real} would need two steps: Firstly, the homogeneous transformation matrix A_i would be obtained by the coordinate frame f_i^{ideal} with respect to f_{i-1}^{real}; Secondly, the homogeneous transformation matrix E_i would be obtained by the coordinate frame f_i^{real} with respect to f_i^{ideal}. Equation (2) follows:

$$E_i = Rot\,(x_i, \varepsilon_{i4})\,Rot\,(y_i, \varepsilon_{i5})\,Rot\,(z_i, \varepsilon_{i6})\,Trans\,(\varepsilon_{i1}, \varepsilon_{i2}, \varepsilon_{i3}) \tag{2}$$

where the three parameters $\varepsilon_{i1}, \varepsilon_{i2}, \varepsilon_{i3}$ indicate the translation values from the origin O_i^{ξ} to $O_i^{\Re}$ in the frame f_i^{ideal} along the X, Y and Z axes, respectively. The other three parameters $\varepsilon_{i4}, \varepsilon_{i5}, \varepsilon_{i6}$ represent the Euler angles of the coordinate frame f_i^{real} with respect to f_i^{ideal} in the Figure 2.

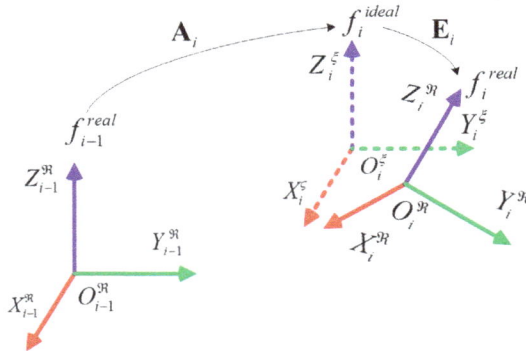

Figure 1. Frame translation and rotation due to the errors for the ith link.

In the Equation (2), the six parameters $\varepsilon_{i1}, \varepsilon_{i2}, \varepsilon_{i3}, \varepsilon_{i4}, \varepsilon_{i5}, \varepsilon_{i6}$ are generally called the generalized geometric error parameters. To simplify the calculation process, the matrix E_i is approximately represented by the Taylor's formula expansion of the Equation (2). The first order values of expansion formula only remain because the coordinate frame f_i^{real} slightly deviates from f_i^{ideal}. Therefore, the matrix E_i is rewritten as the Equation (3):

$$E_i = \begin{bmatrix} 1 & -\varepsilon_{i6} & \varepsilon_{i4} & \varepsilon_{i1} \\ \varepsilon_{i6} & 1 & -\varepsilon_{i5} & \varepsilon_{i2} \\ -\varepsilon_{i4} & \varepsilon_{i5} & 1 & \varepsilon_{i3} \\ 0 & 0 & 0 & 1 \end{bmatrix} \tag{3}$$

The matrix B_i called the generalized geometric error matrix and indicates the transformation matrix of the coordinate frame f_i^{real} with respect to f_{i-1}^{real}.

$$B_i = A_i E_i \tag{4}$$

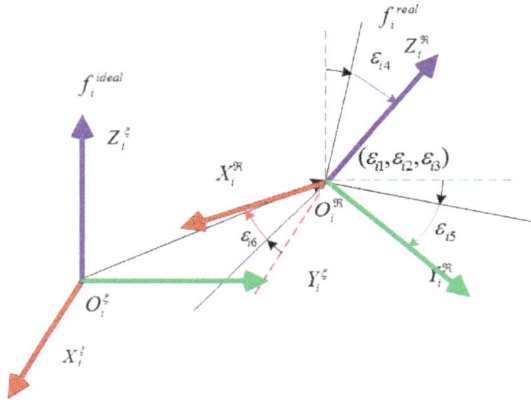

Figure 2. Definition of the generalized geometric error parameter for the ith link.

2.3. Kinematic Model Based on Generalized Geometric Error Theory

There are two steps for establishing the kinematic model of PACMM. Firstly, each joint variable of PACMM is nominally equal to zero at the initial state. In other words, this state is also called the zero pose. Secondly, the kinematic model of PACMM would be established by the generalized geometric error matrix at the initial state. In Figure 3, the homogeneous matrix A_i is obtained by the coordinate frame $O_{i-1}^{\Re} X_{i-1}^{\Re} Y_{i-1}^{\Re} Z_{i-1}^{\Re}$ transformed to $O_i^{\zeta} X_i^{\zeta} Y_i^{\zeta} Z_i^{\zeta}$, and the matrix E_i represents the coordinate frame $O_i^{\Re} X_i^{\Re} Y_i^{\Re} Z_i^{\Re}$ slightly deviating from the ideal frame $O_i^{\zeta} X_i^{\zeta} Y_i^{\zeta} Z_i^{\zeta}$. When $i = 0, 7$ the corresponding frames are the base frame and the center of the rigid probe of PACMM, respectively. It is worth noting that the coordinate frame is not directly established by the D-H model method because there is not the rotation joint at the coordinate frame $O_6^{\Re} X_6^{\Re} Y_6^{\Re} Z_6^{\Re}$. The point P indicates the center of the probe here. Therefore, the coordinate frame is established by the following method: The projective point P' is obtained by the point P projective to the plane of $X_5^{\Re}O_5^{\Re}Z_5^{\Re}$, and the line PP' is in line with the $Z_6^{\Re}$ axis of the coordinate frame. Therefore, the coordinate frame $O_6^{\Re}X_6^{\Re}Y_6^{\Re}Z_6^{\Re}$ is established by the above method in Section 2.2. The coordinate frame $O_7^{\Re}X_7^{\Re}Y_7^{\Re}Z_7^{\Re}$ is obtained by the translation d_7 along the axis in the coordinate frame $O_6^{\Re}X_6^{\Re}Y_6^{\Re}Z_6^{\Re}$. However, the angular encoder includes zero position error at the zero position because of the manufacture and assembly errors of the angular encoder. θ_{i0} is supposed to represent the zero position error of the joint variable θ_i at the initial zero position. Then, the actual joint variable is indicated by $\Theta_i = \theta_{i0} + \theta_i$. The transformation matrix A_i is rewritten as the Equation (5):

$$A_i = \begin{bmatrix} cos\Theta_i & -sin\Theta_i cos\alpha_i & sin\Theta_i sin\alpha_i & l_i cos\Theta_i \\ sin\Theta_i & cos\Theta_i cos\alpha_i & -cos\Theta_i sin\alpha_i & l_i sin\Theta_i \\ 0 & sin\alpha_i & cos\alpha_i & d_i \\ 0 & 0 & 0 & 1 \end{bmatrix} \qquad (5)$$

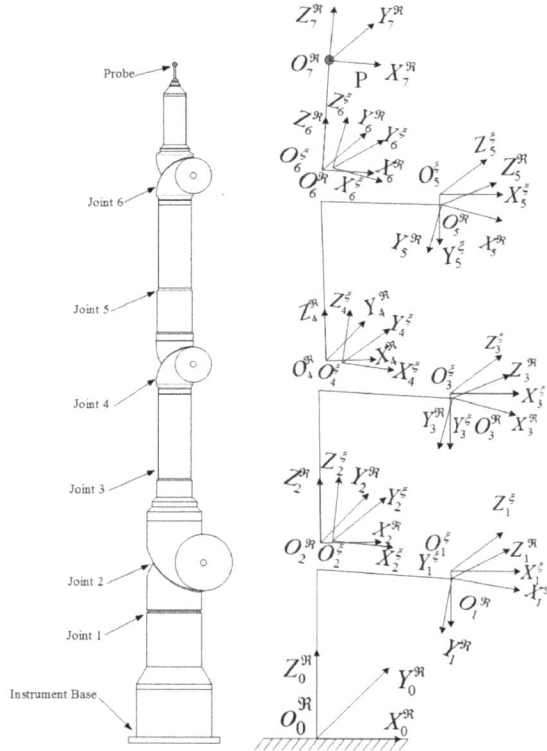

Figure 3. Configuration of portable articulated coordinate measuring machine (PACMM) based on the generalized geometric error model.

Therefore, the relationship matrix T from the base frame $O_0^{\Re} X_0^{\Re} Y_0^{\Re} Z_0^{\Re}$ to the probe frame $O_7^{\Re} X_7^{\Re} Y_7^{\Re} Z_7^{\Re}$ is represented by the Equation (6):

$$T = \prod_i^6 B_i \, Trans\,(0,0,d_7) = \prod_i^6 A_i E_i \, Trans\,(0,0,d_7) \tag{6}$$

where all the generalized geometric error parameters are indicated by using vector $\varepsilon = \left[\varepsilon_{11}, \varepsilon_{12}, \cdots, \varepsilon_{ij}, \cdots, \varepsilon_{66}\right]$ $(i,j = 1, 2, \cdots, 6)$ in the actual kinematic model of PACMM.

3. Error Model and Calibration Algorithm

3.1. Error Model

To ensure the high accuracy measurement results of PACMM, the generalized geometric error parameters vector of Equation (6) must be accurately calibrated except for the nominal values $\theta_{i0}, \alpha_i, l_i, d_i$ $(i = 1, 2, \cdots, 6)$ and d_7. In this paper, an Invar length gauge which has the coefficient of low-thermal expansion is employed as the gauge for calibrating the generalized geometric error parameters of PACMM. For ease of use the Invar length gauge is placed on the three DOFs support platform shown in Figure 4. The pose of the Invar length gauge with regard to PACMM is changed by rotating the three DOFs support platform to collect six joint angle values of PACMM in different poses. The Invar length gauge has three cone-shaped holes on one surface, the length between cone-shaped hole 1 and 2 is $L_{12} = 514.371$ mm and the length between cone-shaped hole 1 and 3 is $L_{13} = 1015.962$ mm. In this paper, L_{12} is selected as the standard value for calibrating the generalized geometric error parameters of PACMM. Therefore, the error model of the generalized geometric error parameters of PACMM is established based on the spatial distance. The measuring points P_k^1 and P_k^2 are obtained when the probe is placed in cone-shaped hole 1 and 2, respectively. Therefore, the error model of PACMM is indicated by Equation (7):

$$E_k\,(\varepsilon) = L_{12} - \|P_k^1 - P_k^2\| \tag{7}$$

where the symbol $\|\bullet\|$ indicates the norm of the vector. k indicates the serial number of measurement points.

According to the least square principle, the objective function is established by Equation (8):

$$F\,(\varepsilon) = \sum_{k=1}^m E_k^2\,(\varepsilon) = \sum_{k=1}^m \left(L_{12} - \|P_k^1 - P_k^2\|\right)^2 \tag{8}$$

where m indicates the number of measurement points.

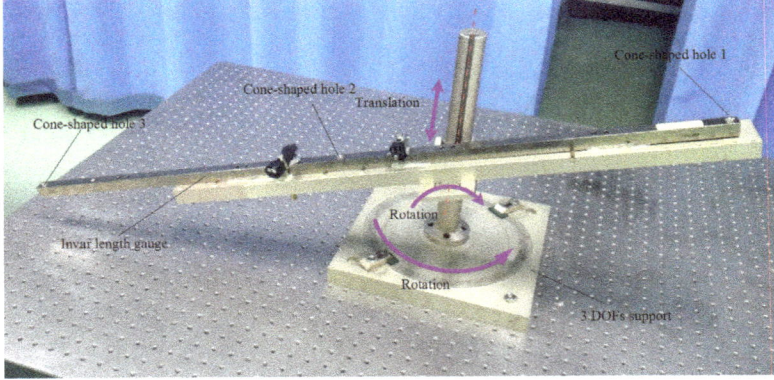

Figure 4. Length gauge.

3.2. Calibration Algorithm

Calibration algorithm is supposed to as the main work for solving the generalized geometric error parameters in Equation (8). The selection of calibration algorithm lies in its convergence rate and identification efficiency. Although GA, PSO, modified SA and nonlinear least square method have been applied in calibrating the kinematic parameters of robot or PACMM, the above-mentioned calibration algorithms with respect to LM algorithm have slower convergence rate and poorer stability. Therefore, LM algorithm is selected as the calibration algorithm for calibrating the generalized geometric error parameters of PACMM. Its detailed calculation procedures are as follows:

Step 1: Set the initial estimation values vector $\varepsilon^{(0)}$ of generalized geometric error parameters vector ε, the initial damping factor $\mu = \mu_0^{(0)}$, permissible error $\epsilon > 0$, set the flag of the iteration step $t = 0$ and the growth factor $v > 0$, $v = 2, 5$ or 10;

Step 2: Calculate $E^{(t)} = E(\varepsilon^{(t)})$, $S^{(t)} = E^{(t)T}E^{(t)}$, $J^{(t)} = \left[\dfrac{\partial E_k^{(t)}(\varepsilon^{(t)})}{\partial \varepsilon_{ij}}\right]$, $(t = 1, 2, \cdots, n)$;

Step 3: Solve $(J^{(t)T}J^{(t)} + \mu^{(t)}I)\Delta\varepsilon^{(t)} = -J^{(t)T}E^{(t)}$;

Step 4: Calculate $\varepsilon^{(t+1)} = \varepsilon^{(t)} + \Delta\varepsilon^{(t)}$, $E^{(t+1)} = E(\varepsilon^{(t+1)})$, $S^{(t+1)} = E^{(t+1)T}E^{(t+1)}$;

Step 5: If $\|\Delta\varepsilon^{(t)}\| < \epsilon$, vector $\varepsilon^{(t)}$ is supposed to as the best estimate ε^*, else $t = t + 1$;

Step 6: If $S^{(t+1)} < S^{(t)}$, then $\mu^{(t+1)} = \mu^{(t)}/v$, go to Step 2, else $\mu^{(t+1)} = \mu^{(t)}v$, go to Step 3.

However, some generalized geometric error parameters resulting in the tiny contributions for the probe position errors of PACMM could not be calibrated. Therefore, these parameters are assimed to be the redundant parameters of generalized geometric error parameters for error model of PACMM. The redundant parameters $\varepsilon_{21}, \varepsilon_{31}, \varepsilon_{32}, \varepsilon_{34}, \varepsilon_{35}, \varepsilon_{41}, \varepsilon_{46}, \varepsilon_{51}, \varepsilon_{52}, \varepsilon_{66}$ are eliminated by using the analysis method introduced by Zhang *et al.* [30].

4. Sample Strategy

In fact, The sampling process for calibration data of PACMM records its six joint values. The calibration accuracy of PACMM must be affected by the sampling randomness in the whole work volume of PACMM. Borm *et al.* [31] introduced an optimized sampling strategy for calibrating the serial robot, needs the joint values of the serial robot collected uniformly in the corresponding joint spaces. Therefore, the length gauge should be uniformly distributed in the whole work volume of PACMM for sampling the calibration data to reduce the sampling randomness of the PACMM. However, the partial work volume of PACMM can not be collected according to performance tests of PACMM. The length gauge is uniformly distributed at four different positions around the PACMM. The length gauge can be uniformly rotated by the different poses to reduce the geometric and non-geometric errors of PACMM according to the Figure 5.

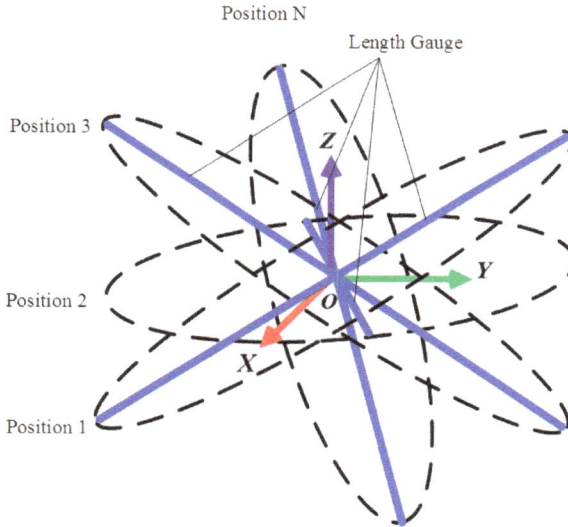

Figure 5. Sample strategy.

The detailed sampling procedures are shown in the Figure 6.

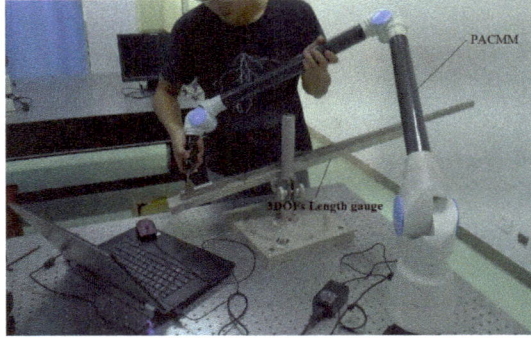

Figure 6. Sampling procedure.

5. Experiment

In this section, the generalized geometric error parameters calibration of PACMM is performed by L-M algorithm in the Section 3.2. In order to verify the advantage of the proposed method with respect to D-H model [32], the comparison experiment is also done by PACMM based on the generalized geometric error parameters and D-H model, respectively.

5.1. Calibration

The nominal values of the kinematic structural parameters of the kinematic model of PACMM are listed in the Table 1. The parameter d_7 of the probe is equal to 210 mm here.

Table 1. Nominal values of structural parameters.

Parameters	A_1	A_2	A_3	A_4	A_5	A_6
θ (rad)	0	−1.57	−3.33	−4.57	−0.97	−3.96
α (rad)	$-\pi/2$	$\pi/2$	$-\pi/2$	$\pi/2$	$-\pi/2$	$\pi/2$
l (mm)	40	−40	32	−32	32	−32
d (mm)	0	0	440	0	340	0

In the error model of PACMM, the initial generalized geometric error parameters vector $\varepsilon^{(0)}$ is equal to 0. In the process of the calibration, the length gauge would be placed by the above method in the Section 4. The calibration data are substituted into the L-M algorithm of the Section 3.2. The convergence value of the objective function $F(\varepsilon)$ is 0.1402 by 50 iterations. Figure 7 shows that the value of $F(\varepsilon)$ tends to be stabilized in about five iterations. The generalized geometric error parameter vector ε^* is supposed to as the optimal parameter vector for PACMM. The generalized geometric error parameters calibrated of PACMM are listed in the Table 2.

52

Table 2. Generalized geometric error parameters calibrated.

Parameters	E_1	E_2	E_3	E_4	E_5	E_6
ε_1 (mm)	−0.0156	−0.0060	−0.0062	−0.0076	0.0164	−0.0038
ε_2 (mm)	-	0.0071	0.0081	−0.0025	0.0204	−0.0017
ε_3 (mm)	-	-	-	-	-	0.0247
ε_4 (rad)	-	0.0015	−0.0307	0.0443	−0.0332	-
ε_5 (rad)	-	0.0288	−0.0423	−0.0318	−0.0013	-
ε_6 (rad)	−0.0011	−0.0021	0.0021	−0.0007	0.0018	-

- Redundant parameter.

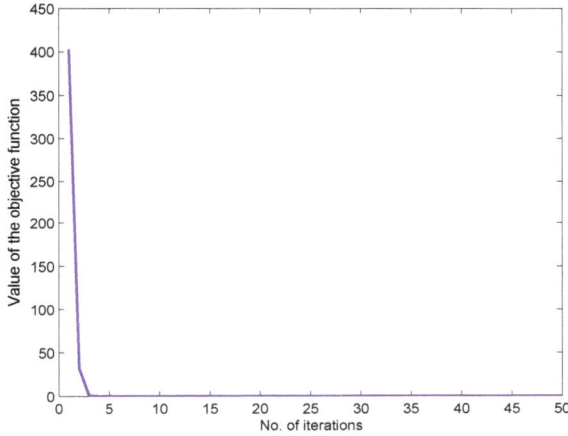

Figure 7. Convergence curve.

The measurement errors of the length gauge using PACMM before and after calibrating 50 times are shown in the Figure 8. In addition, the measurement standard deviation of PACMM for the length gauge is reduced from 0.5550 mm (before calibration) to 0.0452 mm (after calibration).

5.2. Comparison

To verify the advantage of the kinematic model of PACMM based on generalized geometric error model with respect to D-H model, this paper presents that PACMM measures the length gauge 20 times at four different positions and the measured angle data is calculated by generalized geometric error parameters and D-H method, respectively. Figure 9 shows the measurement standard deviations of length gauge using PACMM based on D-H and generalized geometric error method at four different positions. Besides, the results of the comparison experiment demonstrate that the measurement standard deviation of length gauge using PACMM based on the generalized geometric error method is reduced from 0.0627 mm to 0.0452 mm with respect to the D-H model.

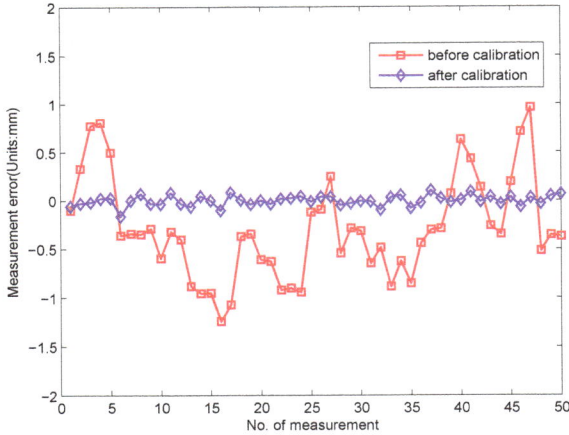

Figure 8. Measurement residual errors of length gauge using PACMM before and after calibration.

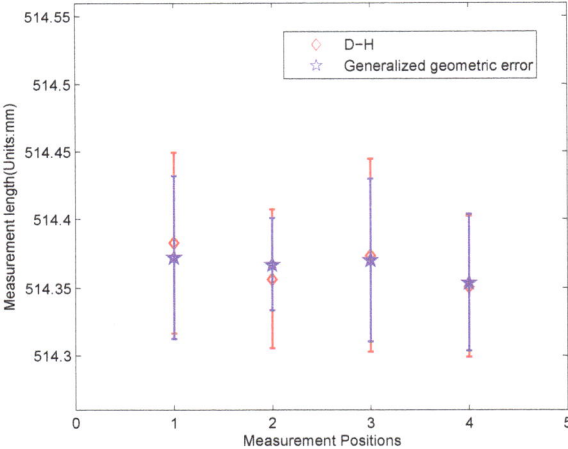

Figure 9. Comparison of the measurement results of length gauge using PACMM based on the Denavit-Hartenberg (D-H) and generalized geometric error method.

6. Discussion

The comparison experimental results demonstrate that the proposed method is effective and the higher measurement accuracy of PACMM using the proposed method could be achieved with respect to D-H model. Note here that the measurement accuracy of FaroArm Platinum 1.8 m is 0.037 mm, as specified by ASME B89.4.22 norm. The measurement accuracy of PACMM using the proposed method is slightly lower than Faro Platinum 1.8 m, but its price is only half of the FaroArm Platinum 1.8 m. In future work, improving the manufacturing process of parts, optimizing the calibration method of the structural parameters, reducing the number of PACMM DOFs in the process of the practical work and adopting multi-group kinematic parameters calibrated for the kinematic model of PACMM would be regarded as the research key to improving the measurement accuracy of PACMM.

7. Conclusions

A new kinematic model of PACMM which considers both geometric and non-geometric errors is proposed on the basis of generalized geometric error model in this paper. Generalized geometric error parameters are calibrated by the L-M algorithm. The experimental results demonstrate that the measurement standard deviation of the length gauge is reduced from 0.5550 mm (before calibration) to 0.0452 mm (after calibration). Besides, the results of the comparison experiment which is also performed to illustrate the advantage of PACMM based on the generalized geometric error method, demonstrate the measurement standard deviation of the proposed method, which is reduced from 0.0627 mm to 0.0452 mm with respect to D-H.

Acknowledgments: This research is supported by National Natural Science Foundation of China (Grant No.: 51375137) and Project 111 (Grant No.: B12019) and National Key Scientific Apparatus Development Project (Grant No.: 2013YQ220893). The authors would like to thank all other members of the research team for their contributions to this research.

Author Contributions: Hui-ning Zhao, Lian-dong Yu and Wei-shi Li conceived and designed the experiments and wrote the paper; Hua-kun Jia performed the experiments; Hui-ning Zhao and Hua-kun Jia analyzed the data; Jing-qi Sun created the figures.

Conflicts of Interest: The authors declare no conflict of interest.

Abbreviations

The following abbreviations are used in this manuscript:

PACMM	Portable Articulated Coordinate Measuring Machine
CMM	Coordinate Measuring Machine
D-H	Denavit-Hartenberg
L-M	Levenberg-Marquard
1D	one-dimensional
DOF	degree of freedom
GA	Genetic Algorithm
PSO	Particle Swarm Optimization
SA	Simulated Annealing

References

1. Yu, L.D.; Zhao, H.N.; Zhang, W.; Li, W.S.; Deng, H.X.; Song, Y.T.; Gu, Y.Q. Development of precision measurement network of experimental advanced superconducting tokamak. *Opt. Eng.* **2014**, *52*, 26–31.
2. Joubair, A.; Slamani, M.; Bonev, I.A. A novel *XY*-Theta precision table and a geometric procedure for its kinematic calibration. *Robot. Comput.-Integr. Manuf.* **2012**, *28*, 57–65.
3. Chen, I.M.; Yang, G.; Tan, C.T.; Yeo, S.H. Local POE model for robot kinematic calibration. *Mech. Mach. Theory* **2001**, *36*, 1215–1239.
4. Denavit, J.; Hartenberg, R.S. A kinematic notation for lower-pair mechanisms based on matrices. *ASME J. Appl. Mech.* **1955**, *22*, 215–221.
5. Hayati, S.A. Robot arm geometric link parameter estimation. In Proceedings of the 22nd IEEE Conference on Decision and Control, San Antonio, TX, USA, 14–16 December 1983; pp. 1477–1483.
6. Judd, R.P.; Knasinski, A.B. A technique to calibrate industrial robots with experimental verification. *IEEE Trans. Robot. Autom.* **1990**, *6*, 20–30.
7. Veitschegger, W.K.; Wu, C.H. Robot calibration and compensation. *IEEE J. Robot. Autom.* **1988**, *4*, 643–656.
8. Zhuang, H.; Roth, Z.S.; Hamano, F. A complete and parametrically continuous kinematic model for robot manipulators. *IEEE Trans. Robot. Autom.* **1992**, *8*, 451–463.
9. Roth, Z.S.; Mooring, B.; Ravani, B. An overview of robot calibration. *IEEE J. Robot. Autom.* **1987**, *3*, 377–385.
10. Meggiolaro, M.A.; Dubowsky, S.; Mavroidis, C. Geometric and elastic error calibration of a high accuracy patient positioning system. *Mech. Mach. Theory* **2005**, *40*, 415–427.
11. Tian, W.; Gao, W.; Zhang, D.; Huang, T. A general approach for error modeling of machine tools. *Int. J. Mach. Tools Manuf.* **2014**, *79*, 17–23.
12. American Society of Mechanical Engineering (ASME). *Methods for Performance Evaluation of Articulated Arm Coordinate Measuring Machines*; AMSE B89.4.22; American Society of Mechanical Engineering: New York, NY, USA, 2005; pp. 1–45.
13. Verein Deutscher Ingenieure (VDI). *Acceptance and Reverification Test for Articulated Arm Coordinate Measuring Machines*; VDI/VDE 2617 Part 9; Verein Deutscher Ingenieure: Düsseldorf, Germany, 2009; pp. 1–20.

14. International Organization for Standardization. *Geometrical Product Specifications (GPS)—Acceptance and Reverification Tests for Coordinate Measuring Systems (CMS)—Part 12: Articulated Arm Coordinate Measurement Machines (CMM)*; ISO/CD 10360-12; ISO: Geneva, Switzerland, 2014; pp. 1–42.

15. Furutani, R.; Shimojima, K.; Takamasu, K. Kinematical calibration of articulated CMM using multiple simple artifacts. In Proceedings of the 17th IMEKO World Congress, Dubrovnik, Croatia, 22–27 June 2003; pp. 1789–1801.

16. Santolaria, J.; Aguilar, J.J.; Yagüe, J.A.; Pastor, J. Kinematic parameter estimation technique for calibration and repeatability improvement of articulated arm coordinate measuring machines. *Precis. Eng.* **2008**, *32*, 251–268.

17. Nguyen, H.N.; Zhou, J.; Kang, H.J. A new full pose measurement method for robot calibration. *Sensors* **2013**, *13*, 9132–9147.

18. Kovač, I.; Frank, A. Testing and calibration of coordinate measuring arms. *Precis. Eng.* **2001**, *25*, 90–99.

19. Shimojima, K.; Furutani, R.; Takamasu, K.; Araki, K. The estimation method of uncertainty of articulated coordinate measuring. In Proceedings of the 2002 IEEE International Conference on Industrial Technology (IEEE ICIT '02), Bangkok, Thailand, 11–14 December 2002; Volume 1, pp. 411–415.

20. Piratelli-Filho, A.; Lesnau, G.R. Virtual spheres gauge for coordinate measuring arms performance test. *Measurement* **2010**, *43*, 236–244.

21. González-Madruga, D.; Cuesta, E.; Patiño, H.; Barreiro, J.; Martinez-Pellitero, S. Evaluation of AACMM Using the Virtual Circles Method. *Procedia Eng.* **2013**, *63*, 243–251.

22. Acero, R.; Brau, A.; Santolaria, J.; Pueo, M. Verification of an articulated arm coordinate measuring machine using a laser tracker as reference equipment and an indexed metrology platform. *Measurement* **2015**, *69*, 52–63.

23. Li, J.; Yu, L.D.; Sun, J.Q.; Xia, H.J. A Kinematic Model for Parallel-Joint Coordinate Measuring Machine. *J. Mech. Robot.* **2103**, *5*, 044501.

24. Light, T.V.; Gorlach, I.A.; Schönberg, A.; Schmitt, R. Measuring arm calibration. In Proceedings of the 5th European Conference on European Computing Conference, Paris, France, 28–30 April 2011; pp. 222–227.

25. Dong, Z.; Zhang, W.; Zhao, H.N.; Yu, L.D. Structural parameter calibration for parallel dual-joint coordinate measuring machine. *Nanotechnol. Precis. Eng.* **2015**, *13*, 287–292.

26. Gao, G.B.; Wen, W.; Lin, K.; Chen Z.G. Parameter identification based on modified annealing algorithm for articulated arm CMMs. *Opt. Precis. Eng.* **2009**, *17*, 2499–2505.

27. Liu, W.L.; Qu X.H.; Yan, Y.G. Self-Calibration and Error Compensation of Flexible Coordinate Measuring Robot. In Proceedings of the 2007 International Conference on Mechatronics and Automation, Harbin, China, 5–8 August 2007; pp. 2489–2494.

28. Levenberg, K. A Method for the Solution of Certain Non-linear Problems in Least Squares. *Quart. Appl. Math.* **1944**, *2*, 164–168.

29. Marquardt, D.W. An Algorithm for Least Square Estimation of Non-Linear. *J. Soc. Ind. Appl. Math.* **1963**, *11*, 431–441.

30. Zhang, T.; Liang, D.; Dai, X. Test of Robot Distance Error and Compensation of Kinematic Full Parameters. *Adv. Mech. Eng.* **2014**, *6*, 1–9.
31. Borm, J.H.; Menq, C.H. Determination of Optimal Measurement Configurations for Robot Calibration Based on Observability Measure. *Int. J. Robot. Res.* **1991**, *10*, 51–63.
32. Zheng, D.T.; Xiao, Z.Y.; Xia, X. Multiple Measurement Models of Articulated Arm Coordinate Measuring Machines. *Chin. J. Mech. Eng.* **2015**, *28*, 994–998.

Methods of In-Process On-Machine Auto-Inspection of Dimensional Error and Auto-Compensation of Tool Wear for Precision Turning

Shih-Ming Wang, Yung-Si Chen, Chun-Yi Lee, Chin-Cheng Yeh and Chun-Chieh Wang

Abstract: The purpose of this study is mainly to develop an information and communication technology (ICT)-based intelligent dimension inspection and tool wear compensation method for precision tuning. With the use of vibration signal processing/characteristics analysis technology combined with ICT, statistical analysis, and diagnosis algorithms, the method can be used to proceed with an on-line dimension inspection and on-machine tool wear auto-compensation for the turning process. Meanwhile, the method can also monitor critical tool life to identify the appropriate time for cutter replacement to reduce machining costs and improve the production efficiency of the turning process. Compared to the traditional ways, the method offers the advantages of requiring less manpower, and having better production efficiency, high tool life, fewer scrap parts, and low costs for inspection instruments. Algorithms and diagnosis threshold values for the detection, cutter wear compensation, and cutter life monitoring were developed. In addition, a bilateral communication module utilizing FANUC Open CNC (computer numerical control) Application Programming Interface (API) Spec was developed for the on-line extraction of instant NC (numerical control) codes for monitoring and transmit commands to CNC controllers for cutter wear compensation. With use of local area networks (LAN) to deliver the detection and correction information, the proposed method was able to remotely control the on-machine monitoring process and upload the machining and inspection data to a remote central platform for further production optimization. The verification experiments were conducted on a turning production line. The results showed that the system provided 93% correction for size inspection and 100% correction for cutter wear compensation.

Reprinted from *Appl. Sci.* Cite as: Wang, S.-M.; Chen, Y.-S.; Lee, C.-Y.; Yeh, C.-C.; Wang, C.-C. Methods of In-Process On-Machine Auto-Inspection of Dimensional Error and Auto-Compensation of Tool Wear for Precision Turning. *Appl. Sci.* **2016**, *6*, 107.

1. Introduction

Precision turning plays an important role in manufacturing industry. The products made by turning process include automotive components, aerospace parts, and precise industrial parts, *etc.* Turning accuracy is mainly influenced by the accuracy of the machine, the condition of the cutter, cutting parameters, and environmental conditions, such as external vibration, the environment temperature, *etc.* Since tool wear could cause more cutting resistance, machining vibration, machining temperature, and machining errors, it usually needs on-line monitoring of the status of tool wear and compensating of the tool wear via offsetting the tool position for the next machining process.

To receive a better performance from a machining process, Greg *et al.* carried out a series of research activities [1,2] on improving the multi-gene genetic programming approach. They proposed a modified multi-gene genetic programming (M-MGGP) method using a stepwise regression approach in which the lower-performance genes were eliminated and the high-performing genes were combined. Validation was carried out by applying surface roughness to modeling when turning hardened American Iron and Steel Institute (AISI) H11 steel, and the results showed that M-MGGP has better performance than that of standard MGGP and other methods. They also proposed a new complexity-based multi-gene genetic programming approach in which the functional relationships between the energy consumption and the input process parameters of a milling process were obtained to find an optimum set of input settings; this will conserve a greater amount of energy from these operations. To improve the generalization ability of MGGP, Greg *et al.* [3] developed a new ensemble-based MGGP framework that used statistical and classification strategies. The method was applied on the reliable experimental database in which the outputs were surface roughness, tool life and power consumption. With the use of an embedded approach to molecular dynamics and MGGP, Greg *et al.* [4] proposed a method to investigate the thermal property of single-layer graphene sheet. In this study, the response of thermal conductivity of the graphene sheet with changes in system temperature and Stone-Thrower-Wales (STW) defect concentration was analyzed. In addition, they [5] also used an explicit model formulated by a molecular dynamics-based computational intelligence approach and a paradigm of a computational intelligence (CI) cluster comprising genetic programming to study the nano-drilling process of boron nitride nanosheet panels.

Much research related to tool wear prediction and monitoring had been carried out in past years. Usui *et al.* [6] proposed a method of cutter wear prediction for certain cutting condition, and conducted verification cutting experiments. With the use of different sensor signals, Dimla *et al.* [7] proposed a monitoring method of cutter wear. Li *et al.* [8] proposed methods for cutter wear inspection and failure diagnosis, which could predict the influence of surface quality of the machined work piece caused by cutter wear. Choi *et al.* [9] developed an intelligent monitoring system which could provide on-line monitoring of the wear condition of the turning tool. Prickett *et al.* studied [10] the ways for monitoring the condition of end mills. With the use of the neural network method, Risbood *et al.* [11] proposed a way to predict the dimension errors caused by cutting force and radial cutting vibration. Panda *et al.* [12] used a back-propagation neural network (BPN) to predict the status of tool wear to avoid cutter breakage.

In order to meet the requirements of tight tolerance and high production yield rates for production, in-process tool wear monitoring and compensation and in-process quality inspection are usually developed and become a part of the manufacturing process. However, because those processes are off-machine or manual, they take more time and incur greater costs due to the additional measurement and inspection instruments and work hours. With the use of the on-line diagnosis method, information and communication technology (ICT), and the empirical statistics method, an on-line monitoring and auto-compensation system for the wear of the turning cutter was developed in this study. The system predicts the machining quality inspection and status of the cutter based on the analysis of on-line turning vibration and machining information while turning is operational. Subsequently, the predicted tool wear will be auto-compensated to the computer numerical control (CNC) controller for next turning work. The method was implemented on an automotive component production line, and both of the original off-machine quality inspection and the manually tool wear compensation were able to be eliminated and replaced by the proposed method.

In the study, preliminary experiments were first conducted to collect the data so that the correlations between the variation of cutting vibration signals, cutter wear, and machining errors could be obtained. The diagnosis algorithms were then developed based on the correlations. Lots of experimental data were collected to analyze the characteristics of tool wear and determine the compensation value for tool wear. Based on the statistical analysis and the developed algorithms, a monitoring system with an auto-compensation function was built. Furthermore, adopting FANUC Open CNC API (application programming interface) Specification provided by FANUC Co., Oshino-mura, Japan, a bilateral communication module for a CNC controller was developed with the ability for on-line communication with the CNC controller in order to extract cutting information (such as instantly executing NC (numerical control) codes, coordinates, and the number of cuts, *etc.*) and send compensation commands to the CNC controller for the predicted tool wear. The module can also save all the monitoring results and information to a remote central computer for continuous engineering improvement in the future. An interface with functions of data acquisition, inspection, and error compensation was developed in C# language for easy operation. Finally, experiments on a CNC turning machine were conducted to verify the feasibility and effectiveness of the proposed system.

2. Preliminary Experiments

An automotive component (Figure 1) (GlobalTek, Taoyuan, Taiwan) made of steel (SAE 1018) was chosen for the experiment. Two finish turning processes for the automotive component were used as an implementation object in this study: (1) T03—finishing for the inner diameter; (2) T04—finishing for the outer diameter. Different tungsten turning tools with Chemical vapor deposition (CVD) coatings of TiCN + Al_2O_3 + TiN were used in the two turning processes. The automotive components was produced on a mass production line. According to the manufacturing requirement, production needs to have a quality inspection of 100%, especially for the inner and outer diameters. Therefore, after the T04 and T03 processes finished, the work pieces were removed from the turning machine for outer/inner-diameter inspection with four air gauge instruments (GlobalTek, Taoyuan, Taiwan). The differences in inspected errors between two consecutive work pieces were regarded as the influence of tool wear, and were used as the tool wear compensation values which were manually inputted in the CNC controller (FANUC, Tokyo, Japan).

Figure 1. Automotive component used in the study.

The preliminary experiments were designed to understand three phenomena: (1) the characteristics of the cutting vibration caused by machining force and tool wear; (2) the correlation between the depth of cut and cutting vibration; (3) the correlation between vibration signals and machining quality. The experimental results were used to design the algorithms for dimension inspection and tool wear compensation. The experimental analysis includes: (1) the correlation between dimension error and cutting vibration; (2) an investigation of the vibration pattern caused by tool wear. In order to carry out on-line collection the cutting vibration signals for analysis, two accelerometers (PCB Piezatronics Inc., Depew, NY, USA) were attached to the tool posts (Figure 2), which are close to the work piece to collect the actual vibration signals.

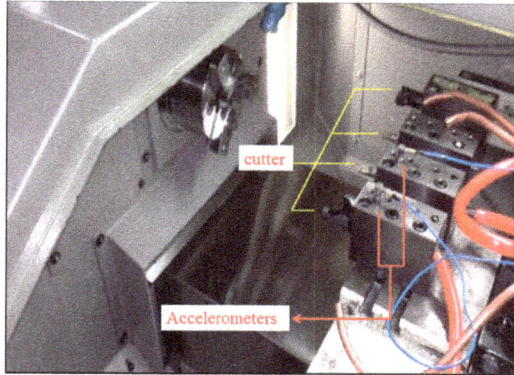

Figure 2. Setup of cutter and accelerometers.

2.1. Correlation between Dimension Error and Cutting Vibration

Tool wear could influence the depth of cut, and the variation of the depth of cut could cause changes to the cutting vibration. Variation in the depth of cut could also cause dimension errors in the component. Using cutting vibration as an index of tool wear, it is very important to ensure that the cutting vibration is sensitive enough to identify the occurrence of tool wear, and sensitive enough to differentiate the variation in the dimension errors.

The experiment started with 0.08 mm for the depth of cut, and increased by 0.01 mm each time to 0.12 mm. The cutting vibration for each cut was measured by the accelerometers, and root-mean-square values of the vibration were calculated. Figures 3 and 4 respectively, show the correlation of the depth of cut and cutting vibration for T03 and T04. It is noted that the cutting vibration is nearly linear in proportion to the depth of cut (0.05 g/0.01 mm), and the depth of cut can be estimated based on the measured vibration of a stable machining process. The same experiment was repeated 14 times, and the results were very repeatable.

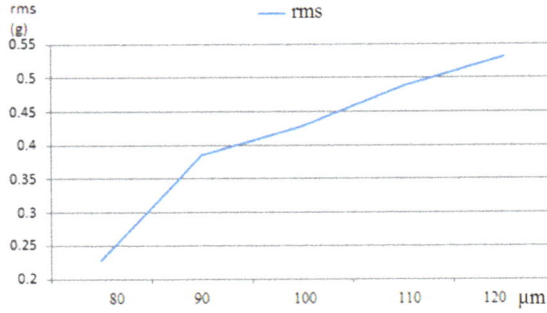

Figure 3. Correlation of depth of cut and cutting vibration for the machining process T03.

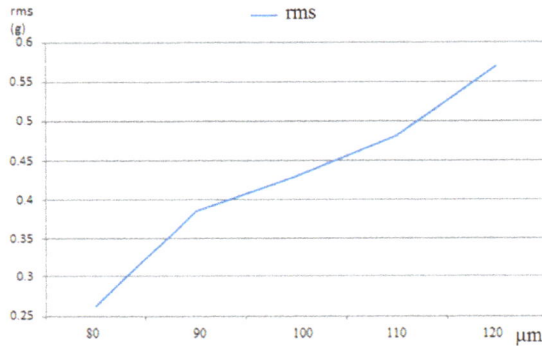

Figure 4. Correlation of depth of cut and cutting vibration for machining process T04.

2.2. Investigation of Tool Wear Patterns and Cutting Vibration Patterns

In a stable machining process, when the number of cuts increases, the tool wear of a cutter increases with a repeatable pattern. Through empirical analysis, the correlation of tool wear and the number of cuts can be obtained and used to build the pattern. Subsequently, the pattern can be used to check whether the tool wear increases normally or not. If the tool wear increases following the pattern, it could be estimated based on the statistical model built with collected experimental data.

In the preliminary experiments, the machining processes T03 and T04 were consecutively conducted, and the relationship between tool wear and its associated numbers of cuts was investigated. The tool wear was measured using a vision-based measurement system. Figures 5 and 6 respectively, show the correlation of tool wear and the numbers of cuts of T03 and T04. It is noted that the tool wear of a cutter have three developing stages: faster wear, normal wear, and critical wear. The fast wear stage usually happens when a new cutter has begun to be used, because the new

64

cutter is sharper and wear occurs more easily. At the normal wear stage, the tool wears slowly and stably. After the normal wear stage, tool wear will increase again at a faster rate. When a cutter is close to its tool life, tool wear will increase rapidly and it will soon to be worn out. This is regarded as critical wear. When critical wear occurs, the cutter wears out quickly and could damage the work piece. Therefore, critical wear could be a significantly signal to monitor for cutter replacement.

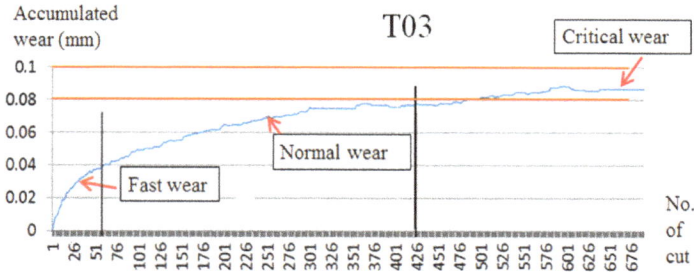

Figure 5. Tool wear *vs.* number of cuts of T03.

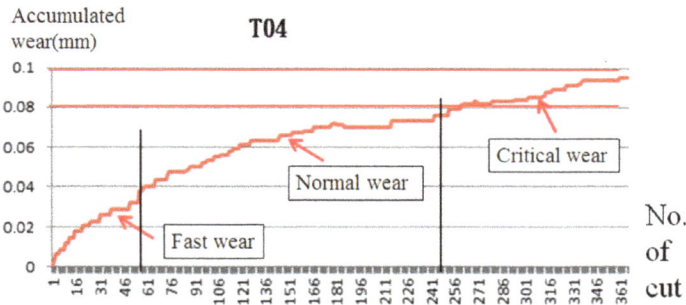

Figure 6. Tool wear *vs.* number of cuts of T04.

From the figures, it was found that the cutter had fast wear when it was new (for example, for the first 50 cuts for T03, and the first 60 cuts for T04), and then the wear increased slowly until the accumulated wear was close to the upper limit of the dimension tolerance of 0.1 mm. As we can see from Figure 5, the tool wear of T03 increased slowly and stably when the accumulated wear was within the range of 40 µm to 77 µm. This period was regarded as the normal wear stage. The total number of cuts made within this range were 425 (from cut 51 to cut 476). The tool wear that happened at this stage was very small (<1 µm). After the 476th cut, the tool wear increased rapidly at about 90 µm for about 100 cuts. Since the required tolerances of the machining are ±10 µm for both the inner diameter and the outer diameter, it is easy to produce a No-Go (*i.e.*, dimension accuracy not qualified) part when the accumulated wear of the cutter approaches 0.1 mm. Therefore, the cutter should be replaced at this time. According to the results, the tool life for T03 and T04 could be respectively defined as 450 cuts and 250 cuts.

Different levels of tool wear will cause different cutting resistances, which will generate different cutting vibrations. In order to use cutting vibration as the index for tool wear monitoring, it is necessary to understand the characteristics of the change in the cutting vibration pattern caused by tool wear. Thus, while conducting the preliminary experiments for the relationship between tool wear and the number of cuts, the associated cutting vibration signals were also on-line measured by the accelerators. Figures 7 and 8 respectively, show the cutting vibration at different cuts of T03 and T03. It is noted that when the cutter showed light wear, the cutting vibration (<2 g) was quite stable (for cut 2 and 250 for T03; cut 2 and 100 for T04). As the tool wear increased, few significant vibrations (4 g) occurred (cut 200 for T03; cut 350 for T04). When the tool wear became greater, significant vibration occurred more frequently (this happened at cut 450 and 500 for T03 and cut 300 and 350 for T04). When comparing Figures 5 and 6 it can be concluded that the significant vibrations were mainly caused by the tool wear. When significant vibration occurred frequently, the tool wear had reached 80 μm and up. According to the results, it can be concluded that the variation of vibration and the frequency of occurrence could be the two indexes for monitoring the status of tool wear. For a stable turning process, the tool wear can be predicted based on the number of cuts. Meanwhile, by checking the cutting vibration pattern, it is able to know whether the cutter has abnormal wear conditions. If a cutter is at the normal wear stage but is experiencing large vibrations, it means the cutter has an abnormal wear status.

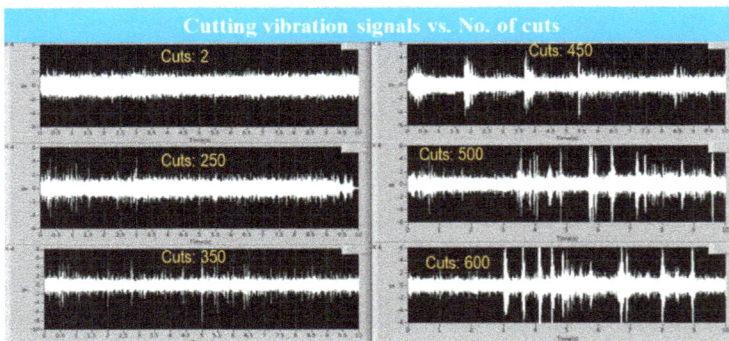

Figure 7. Cutting vibration *vs.* number of cuts for machining process T03.

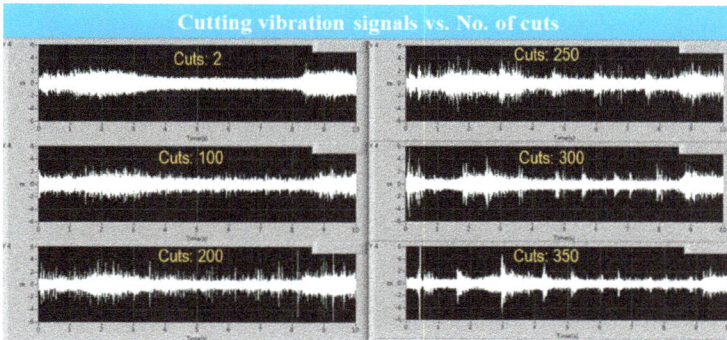

Figure 8. Cutting vibration *vs.* number of cuts for machining process T04.

3. Algorithms for On-Machine Inspection and Tool Wear Auto-Compensation

The on-line inspection algorithm for Go/No-Go (Go: dimension accuracy is qualified; No-Go: dimension accuracy is not qualified) and the algorithm of tool wear auto-compensation were developed based on the statistical analysis of experimental data.

3.1. Algorithm for On-Machine Dimension Error Inspection

For a stable cutting process, cutting force is mainly influenced by the depth of cut, the spindle speed and federate, while different cutting forces cause different cutting vibrations. Tool wear influences the dimension and geometry of a cutter. A worn turning cutter gives a smaller depth of cut, which will cause errors in the machined diameter of the work piece. In addition, different depths of cut cause different cutting forces which influence the cutting vibration. Therefore, it is possible to use cutting vibration to predict tool wear and, furthermore, to estimate the machining error caused by tool wear.

In the experiments, five depths of cut (0.08, 0.09, 0.1, 0.11, 0.12 mm) were chosen for T03 and T04. While turning, the cutting vibrations were measured and their root-mean-square (r.m.s.) values were calculated. Figures 9 and 10 show the correlation between the cutting vibration and the depth of cut. As we can see from the two figures, the cutting vibration was about proportional to the depth of cut. The curves of T03 and T04 are very similar. When the depth of cut increased by 0.01 mm, the cutting vibration increased by about 0.05 g. To ensure the phenomenon shown in Figures 9 and 10 is true and repeatable, 14 experiments were conducted, and all the results were very similar. It implies that the correlation between cutting vibration and depth of can be used to develop the algorithm for on-machine dimension error inspection. The inspection algorithm is designed as follows:

67

(i)   On-line extract the vibration signals (10-s length) of the first three cuts to compute the average r.m.s. value as the reference value.

(ii)   Calculate the maximum allowable error based on the tolerance error given by the customer.

(iii) Convert the maximum allowable error into allowable vibration based on the curves shown in Figures 9 and 10.

(iv)   On-line extract the vibration signals of the new cut. If the r.m.s. value of the cutting vibration of the new cut is greater than the allowable vibration, the machined work piece is a No-Go part. Otherwise, it is a qualified part.

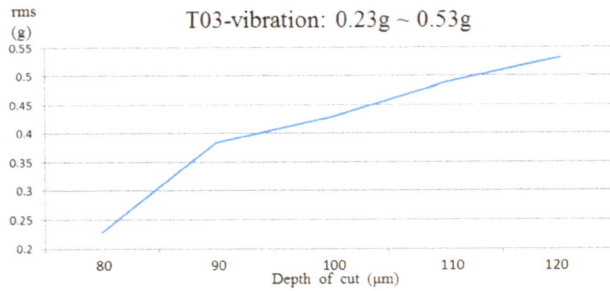

Figure 9. Cutting vibration *vs.* depth of cut for T03.

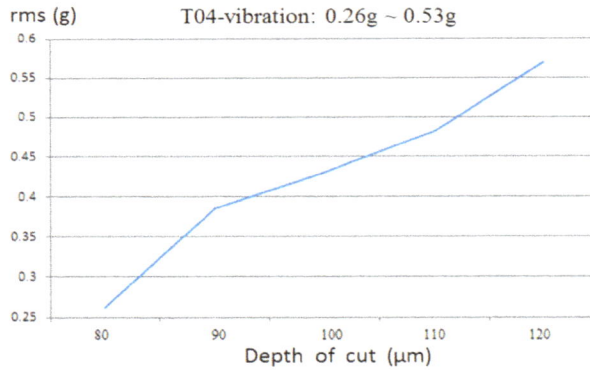

Figure 10. Cutting vibration *vs.* depth of cut for T04.

Figure 11 shows the flowchart of the on-machine dimension error inspection.

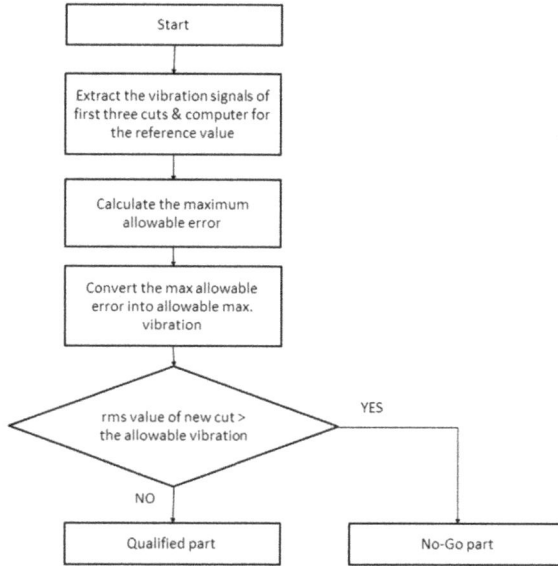

Figure 11. Flowchart of on-machine dimension inspection.

3.2. Algorithm for Prediction and Auto-Compensation of Tool Wear

Tool wear is influenced by cutting resistance, cutting temperature, the geometry of the cutter, the material of the work piece and cutter, the cutting parameters, and the number of cuts, *etc.* Wear of a new cutter increases slowly at the beginning, and will increase rapidly after many cuts are performed. If a machining process is stable and has high repeatability, the tool wear can usually be described with a model developed with regression analysis and the curve fitting method. Thus, it is feasible to use regression analysis and curve fitting methods with the experimental data of tool wear to build an empirical model for tool wear prediction and compensation. With 17 sets of experimental data for each process, Figures 12 and 13 show the statistics of tool wear for T03 and T04, respectively. For T03, after 500 cuts, the accumulated tool wear was about 90 µm. For T04, after 300 cuts, the accumulated tool wear was about 75 µm. In addition, it can be seen that both T03 and T04 show good stability and repeatability in their tool wear trends. For the same number of cuts, the variation of tool wear was 11 µm for T03 and 13 µm for T04, which is smaller than the allowable tolerance of dimension (15 µm). Thus, it makes it possible to build regression models based on the experimental data for tool wear prediction. With the use of the curve fitting method, Equations (1) and (2), which respectively describe the relationship between tool wear and the number of cuts for T03 and T04, were obtained.

$$y = -7 \times 10^{-14}x^6 + 1 \times 10^{-10}x^5 - 9 \times 10^{-8}x^4 + 3 \times 10^{-5}x^3 - 6.6 \times 10^{-3}x^2 + 0.8672x + 6.5828 \quad (1)$$

$$y = -2 \times 10^{-12}x^6 + 2 \times 10^{-9}x^5 - 9 \times 10^{-7}x^4 + 2 \times 10^{-4}x^3 - 0.0192x^2 + 1.2403x + 1.7959 \quad (2)$$

where y represents tool wear (μm), and x represents number of cuts. For a stable turning process, if the number of cuts is known and substituted into Equations (1) and (2), the tool wear can be predicted and used for compensation. The compensation value is directly sent to CNC controller to offset the position of the cutter. Because the predicted wear is an accumulated value, the compensation value should be the difference between the current wear and the previous wear. Based on Equations (1) and (2), Figures 14 and 15 show the tool wear curves (to be used as a compensation reference) for T03 and T04.

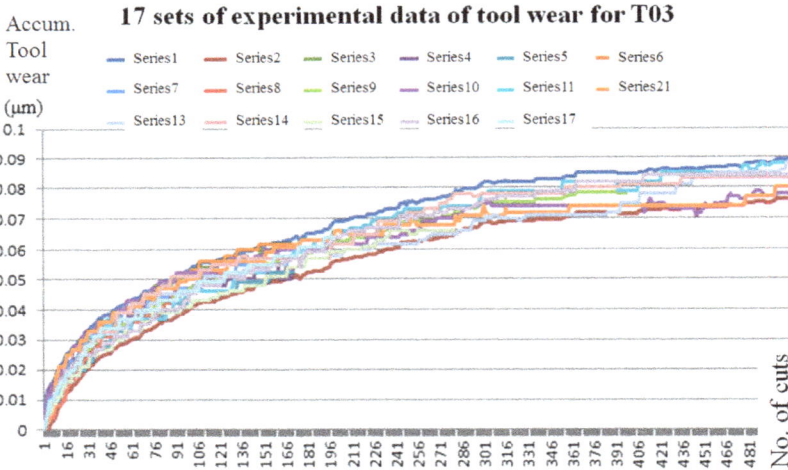

Figure 12. Statistics of tool wear for T03.

The effectiveness of the tool wear compensation also depends on the machine accuracy. If the tool wear is less than the machine's positioning accuracy, the compensation can be skipped until the accumulated wear is greater than the machine's positioning accuracy. Because of it, the compensation strategy is designed based on when the consecutive cuts have very similar tool wear (difference < 1 μm); the accumulated tool wear of those cuts are regarded as identical so that the compensation values calculated based on the tool wear curves for those cuts are zero. The green curves with a step-shape in Figures 14 and 15 represent this compensation strategy.

23 sets of experimental data of tool wear for T04

Accum.
Tool
wear
(μm)

— Series1	— Series2	— Series3	— Series4	— Series5	— Series6
— Series7	— Series8	— Series9	— Series10	— Series11	— Series21
— Series13	— Series14	— Series15	— Series16	— Series17	— Series18
— Series19	— Series20	Series21	— Series22	Series23	

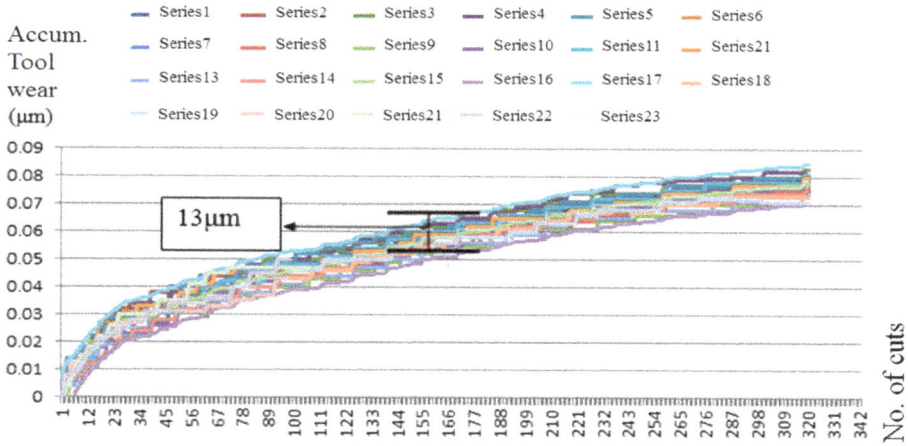

13μm

No. of cuts

Figure 13. Statistics of tool wear for T04.

Accumutated
tool wear(μm)

Tool wear curve based on curve fitted equation — T03

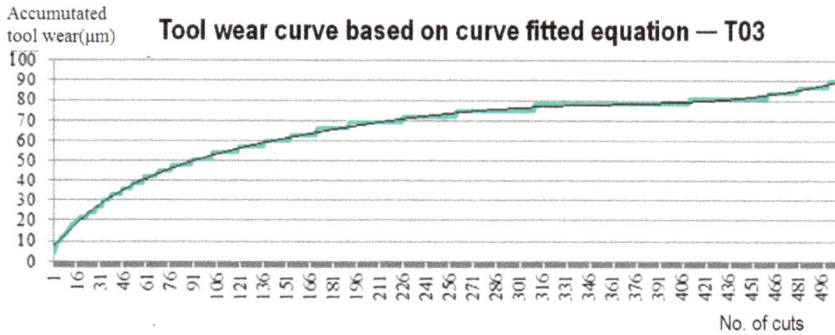

No. of cuts

Figure 14. Tool wear curve for T03.

Accumutated
tool wear(μm)

Tool wear curve based on curve fitted equation — T04

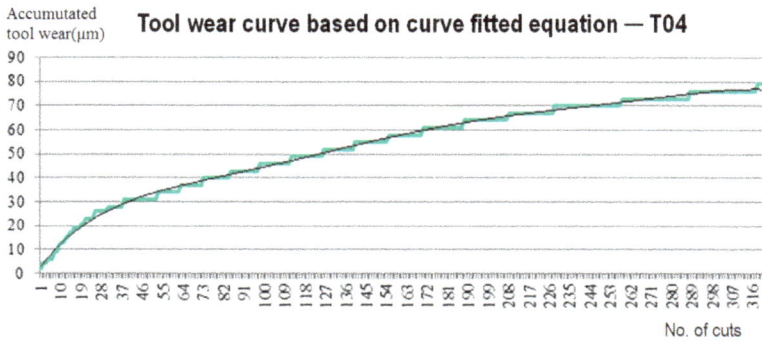

No. of cuts

Figure 15. Tool wear curve for T04.

71

Equations (1) and (2) are good for a stable tool wear condition. Due to the imperfect manufacturing of the cutter or unexpected errors, a cutter may have abnormal wear which is different from the predicted wear solved by Equations (1) and (2). As aforementioned, tool wear could affect the depth of cut. If abnormal tool wear occurs, the cutting vibration patterns collected on-line will be different from that in Figures 7 and 8. Thus, it is necessary to on-line monitor the cutting vibration for tool wear prediction instead of only using Equation (1) and (2). When abnormal tool wear happens, the monitoring system should send an alarm message to operators/engineers to double-check. The algorithms for the prediction and auto-compensation of tool wear are as follows:

(i)   Get the number of cuts from the CNC controller.
(ii)   Input the number of cuts into Equation (1) and (2) to compute for the current tool wear.
(iii)   Calculate the difference between the current tool wear and the previous tool wear, and use it as the compensation value.
(iv)   On-line monitor the cutting vibration, and check whether it is an abnormal vibration.
(v)   If it is not an abnormal vibration, send the compensation value to the CNC controller to offset the cutter for the next machining application; If it is an abnormal vibration, send an alarm message to notify the operator that abnormal tool wear is occurring.

4. Bilateral Communication Module

To directly communicate the CNC controller for the extraction of machining information for tool wear monitoring and auto-compensation, a bilateral communication module that can call the Application Programming Interface (API) to extract instantly executing NC codes, the true spindle speed and feedrate for dimensional error inspection and tool wear monitoring (Figure 16) was developed based on FANUC Open CNC API Spec and Ethernet protocol. The module can also extract other instant machining information and control parameters from the CNC controller. The module enables the proposed system to remote control the on-machine monitoring process and upload the machining and inspection data to the remote central platform for further production optimization. Figure 17a is the human-machine interface containing the on-line extracted information, such as the coordinates of the cutter, the machining parameters, and the CNC control parameters, *etc.* Furthermore, the module can synchronously communicate with several machines for production line monitoring and control. Figure 17b is the screen showing several machines simultaneously connected and monitored.

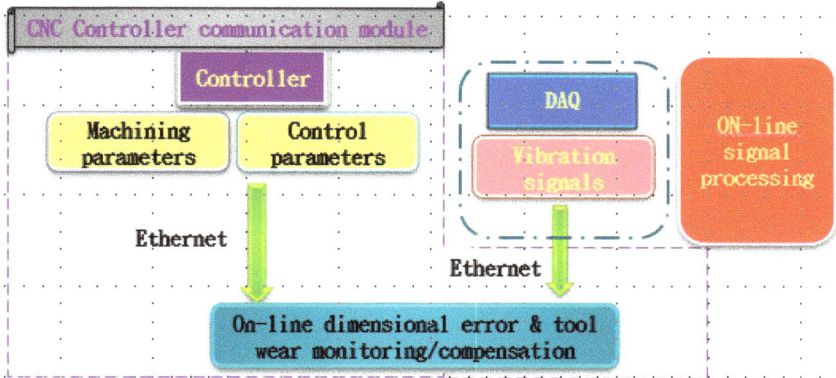

Figure 16. The communication module extracts information for monitoring/compensation.

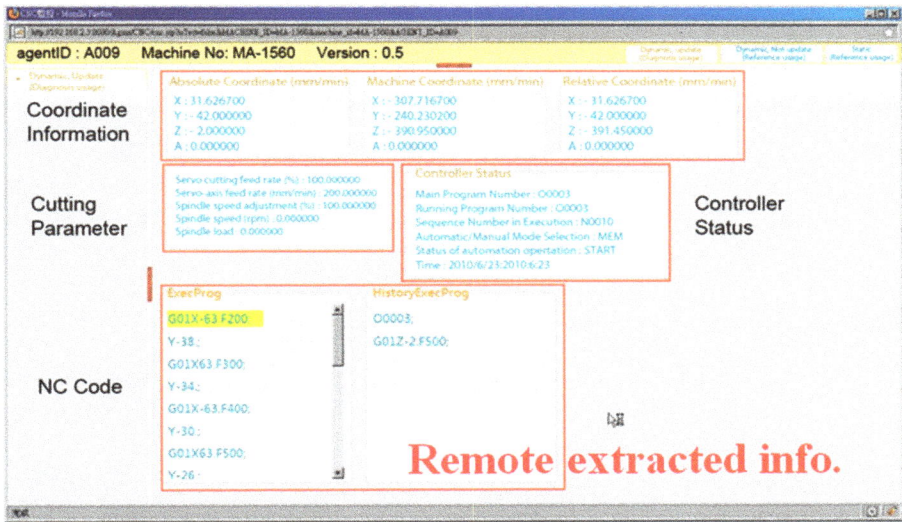

(a)

Figure 17. *Cont.*

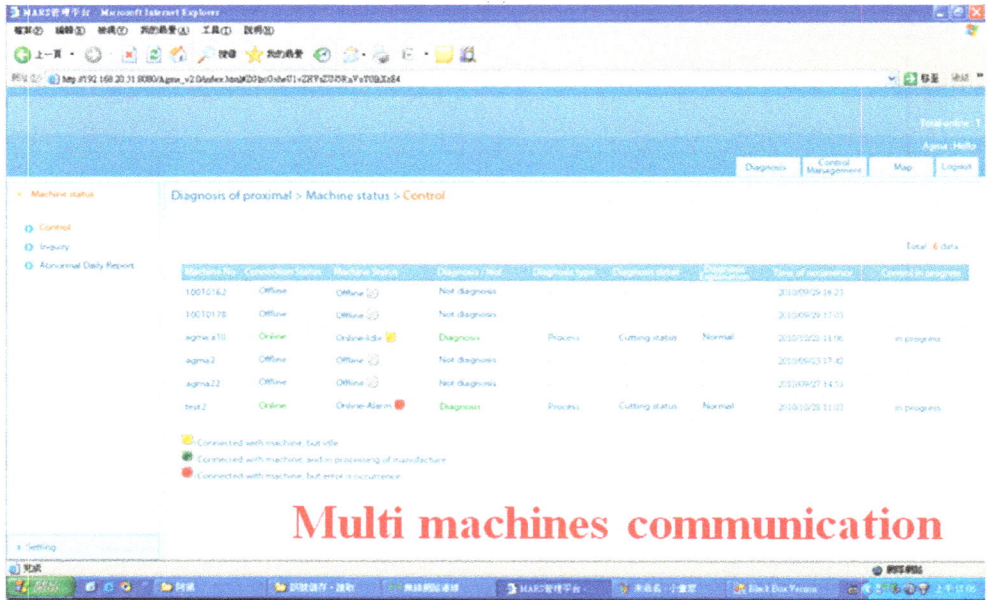

(b)

Figure 17. Bilateral communication module for computer numerical control (CNC) controller. (**a**) On-line extracted information from CNC controller; (**b**) The module synchronously communicates with several machines.

5. Human-Machine Interface

According to the developed algorithms, a system with on-machine inspection functions for machining errors, on-machine monitoring of tool wear, and auto-compensation of tool wear, was built in C# language. To provide a convenient operation, a human-machine interface was built in C# to easily connect to a data acquisition device and a CNC controller for bilateral data transmission. Moreover, it also provides access for executing the proposed functions. Figure 18 shows the developed human-machine interface, and Figure 19 shows the flowchart of the function execution.

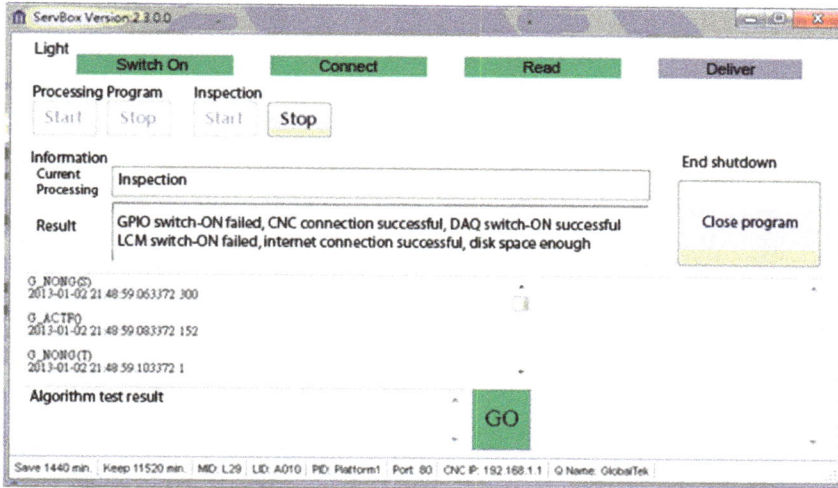

Figure 18. Human-machine interface of the monitoring system.

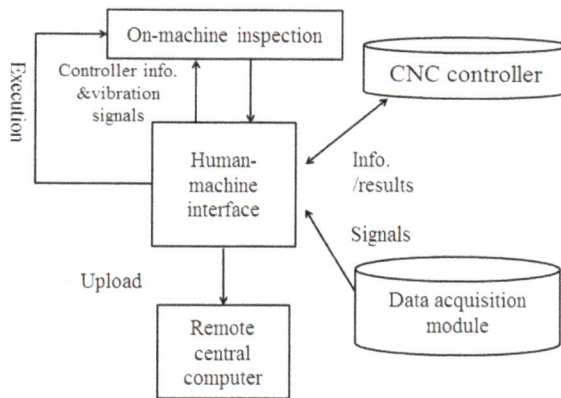

Figure 19. Flowchart of the function execution.

6. Experimental Verification

Two experiments, including on-machine machining error (diameter error) inspection and on-machine tool wear monitoring auto-compensation, were conducted to verify the feasibility and effectiveness of the proposed algorithms and the developed system. The automotive component part shown in Figure 1 was the tested work piece, because the component was produced in a mass production line. The experiments were directly applied to the two finish turning processes for T03 and T04. Table 1 shows the experiment conditions. The tolerance of the inner and outer diameters is ±6 μm. Two accelerometers were attached to the cutter post

75

to on-line collect the cutting vibration signals for monitoring. Meanwhile, machining information, such as instantly executing NC codes, machining parameters, *etc.*, were also extracted with the bilateral communication module as a reference for monitoring.

Table 1. Machining conditions.

Item	Description
Turning cutter	Tungsten Carbide with CVD coating
Material of workpiece	S27 low-carbon steel
Machining type	Inner and out diammeters (T03: inner dia.; T04: outer dia)
Depth of cut	0.1 mm
Cutting length	12 mm
Spindle speed	1500
federate	0.15 mm/rad
Cutting fluid	yes
Tolerance	$\pm$ 6 μm

6.1. On-Machine Machining Error (Diameter Error) Inspection

In this experiment, the on-machine machining error auto-inspection function was applied to the production line of GlobalTek Co. (Taoyuan, Taiwan) to verify the effectiveness of the inspection function. A total of 7643 components were tested in this experiment. The inspection results made by the auto-inspection function were compared to that made using the air gauge instruments, which were originally used by the company. The threshold value of the inspection was set to ±5 μm. Of the produced components, 93% were qualified, and 7% of the produced components were inspected as No-Go parts. The results were very close to the results made using the air gauge instruments. It shows that the proposed function can be used to replace the original manual inspection process made with air gauge instruments. It saved work hours and manufacturing cost for the production. Figure 18 shows the inspection results. Figures 20 and 21 show the statistics of the inspection of T03 and T04, respectively.

Figure 20. On-machine machining inspection result for T03.

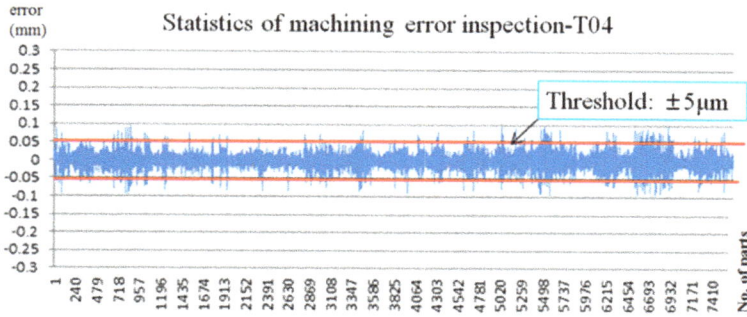

Figure 21. On-machine machining inspection result for T04.

6.2. On-Machine Tool Wear Monitoring and Auto-Compensation

In this experiment, the two functions, on-machine auto-monitoring and auto-compensation of tool wear, were used in the production line of GlobalTek Co. to verify the effectiveness and feasibility of the two functions. Except for the tolerances (T03: ± 6 μm; T04: ± 7.5 μm), the same machining conditions as shown in Table 1 were used. Another 7643 components were tested in this experiment. With use of the proposed system, the tool wear of each cut was predicted and auto-compensated to the CNC controller of the turning machine for the next component.

Figures 22 and 23 show the statistics of the dimension inspection results made with the air gauge instruments. It can be seen that all the machined components matched the requirement for dimension tolerance. It was also noted that most of the parts had dimension errors of about 2–3 μm. Because of the tool wear, some parts had larger dimension errors of about 4–5 μm, but they all still met the tolerance requirement. The results showed that the two functions (on-machine auto-monitoring and auto-compensation of tool wear) can provide appropriate prediction and compensation of tool wear for each cutting process.

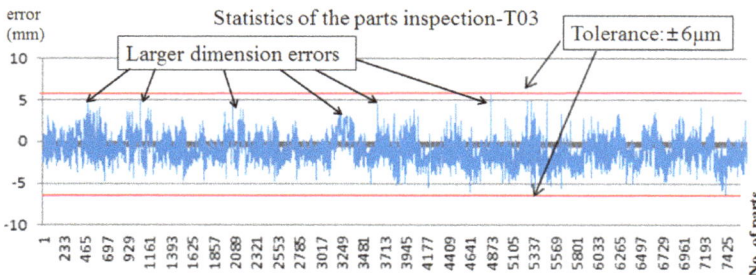

Figure 22. Statistics of parts inspection for T03.

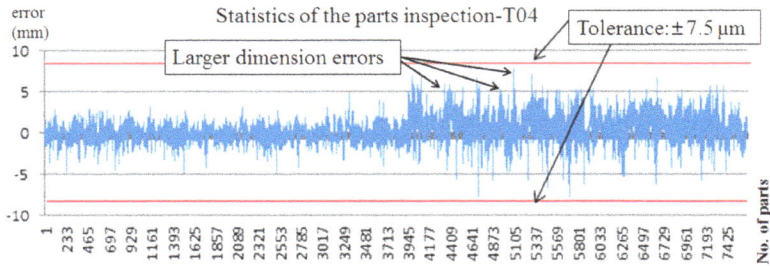

Figure 23. Statistics of parts inspection for T04.

According to the verification results, it showed that the proposed method is suitable for on-line monitoring/compensation for a mass production line. The proposed method is mainly based on the correlation between the variation of cutting vibration signals (detected by the vibration sensors), cutter wear, and machining errors. However, very light tool wear causes very little change in cutting vibration, which it may not be possible to clearly detect using a vibration sensor if the sensitivity of the sensor used is not fine enough. Therefore, the possible limitation of this method is the sensitivity of the selected vibration sensors. The better the sensors, the higher the cost. Thus, the vibration sensors should be selected based on the requirements of the machining process.

7. Conclusions

This study developed an intelligent system with functions of on-machine error inspection and auto-monitoring/auto-compensation of tool wear for precision tuning processes. With the use of vibration signal processing/characteristics analysis technology, combined with ICT, statistical analysis and diagnosis algorithms, the system can proceed with on-line dimension error inspection and on-line tool wear auto-compensation for turning processes. The system can also monitor the critical tool life such that the appropriate time for cutter replacement can be identified to reduce manufacturing costs and improve the production efficiency of a turning process. Algorithms for on-machine dimension error inspection and tool wear auto-compensation were proposed. Based on the algorithms, the system was built in C# language. The results of the verification experiments showed that the developed functions provided correct inspection and appropriate tool wear compensation for turning processes in order to save manufacturing time and costs.

Acknowledgments: This study was supported by Ministry of Science and Technology under the grant number MOST 104-2221-E-033-010 and MOST 104-3011-E-033-001, and by Industrial Technology Research Institute under the program The features of Multi-coupling Dynamic Signal Analysis Technology.

Author Contributions: S.-M.W. conceived and designed the method and algorithm. Y.-S.C. and C.-Y.L. conducted the experiments. Y.-S.C. and C.-C.W. analyzed the data. S.-M.W. and C.-C.Y. wrote the paper.

Conflicts of Interest: The authors declare no conflict of interest.

References

1. Garg, A.; Tai, K. Stepwise approach for the evolution of generalized genetic programming model in prediction of surface finish of the turning process. *Adv. Eng. Software* **2014**, *78*, 16–27.
2. Garg, A.; Lam, J.S.L.; Gao, L. Energy conservation in manufacturing operations: Modelling the milling process by a new complexity-based evolutionary approach. *J. Clean. Prod.* **2015**, *108*, 34–45.
3. Garg, A.; Lam, J.S.L. Improving environmental sustainability by formulation of generalized power consumption models using an ensemble based multi-gene genetic programming approach. *J. Clean. Prod.* **2015**, *102*, 246–263.
4. Garg, A.; Vijayaraghavan, V.; Wong, C.H.; Tai, K.; Gao, L. An embedded simulation approach for modeling the thermal conductivity of 2D nanoscale material. *Simul. Model. Pract. Theory* **2014**, *44*, 1–13.
5. Garg, A.; Vijayaraghavan, V.; Lam, J.S.L.; Singru, P.M.; Gao, L. A molecular simulation based computational intelligence study of a nano-machining process with implications on its environmental performance. *Swarm Evol. Comput.* **2015**, *21*, 54–63.
6. Usui, T.S. Analytical prediction of cutting tool wear. *Wear* **1984**, *100*, 129–151.
7. Dimla, E.; Snr, D. Sensor signals for tool-wear monitoring in metal cutting operations—A review of methods. *Int. J. Mach. Tools Manuf.* **2003**, *40*, 1073–1098.
8. Li, D.; Mathew, J. Tool wear and failure monitoring techniques for turning—A review. *Int. J. Mach. Tools Manuf.* **1990**, *30*, 579–598.
9. Choi, G.S.; Wang, Z.X.; Dornfeld, D.A.; Tsujino, K. Development of an intelligent on-line tool wear monitoring system for turning operations. In Proceedings of the USA-Japan Symposium on Flexible Automation, Asia Pacific Rim Conference ISCIE, Kyoto, Kyoto, Japan, 18–20 June 1990; pp. 683–690.
10. Prickett, P.W.; Johns, C. An overview of approaches to end milling tool monitoring. *Int. J. Mach. Tools Manuf.* **1999**, *39*, 105–122.
11. Risbood, K.A. Prediction of surface roughness and dimensional deviation by measuring cutting forces and vibrations in turning process. *J. Mater. Process. Technol.* **2003**, *132*, 203–214.
12. Panda, S.S.; Singh, A.K.; Chakraborty, D.; Pal, S.K. Drill wear monitoring using back propagation neural network. *J. Mater. Process. Technol.* **2004**, *172*, 283–290.

Development of an Abbe Error Free Micro Coordinate Measuring Machine

Qiangxian Huang, Kui Wu, Chenchen Wang, Ruijun Li, Kuang-Chao Fan and Yetai Fei

Abstract: A micro Coordinate Measuring Machine (CMM) with the measurement volume of 50 mm × 50 mm × 50 mm and measuring accuracy of about 100 nm (2σ) has been developed. In this new micro CMM, an *XYZ* stage, which is driven by three piezo-motors in *X*, *Y* and *Z* directions, can achieve the drive resolution of about 1 nm and the stroke of more than 50 mm. In order to reduce the crosstalk among *X*-, *Y*- and *Z*-stages, a special mechanical structure, which is called co-planar stage, is introduced. The movement of the stage in each direction is detected by a laser interferometer. A contact type of probe is adopted for measurement. The center of the probe ball coincides with the intersection point of the measuring axes of the three laser interferometers. Therefore, the metrological system of the CMM obeys the Abbe principle in three directions and is free from Abbe error. The CMM is placed in an anti-vibration and thermostatic chamber for avoiding the influence of vibration and temperature fluctuation. A series of experimental results show that the measurement uncertainty within 40 mm among *X*, *Y* and *Z* directions is about 100 nm (2σ). The flatness of measuring face of the gauge block is also measured and verified the performance of the developed micro CMM.

Reprinted from *Appl. Sci.* Cite as: Huang, Q.; Wu, K.; Wang, C.; Li, R.; Fan, K.-C.; Fei, Y. Development of an Abbe Error Free Micro Coordinate Measuring Machine. *Appl. Sci.* **2016**, *6*, 97.

1. Introduction

With the fast development of Micro Electro Mechanical Systems (MEMS) and micromachining technology, micro-parts at micro scale are developed and used rapidly, such as micro gears, micro motors and micro sensors, *etc.* These micro-parts have sizes ranging from sub-millimeter to tens of millimeters, and their local geometrical features need to be measured at sub-micrometer accuracy. In many cases, the complete inner and outer geometry of these micro-parts must be verified to ensure their quality and functionality, whereas small geometrical features are inaccessible by conventional Coordinating Measuring Machines (CMMs). In order to measure those microparts, some novel CMMs with accuracy of tens of nanometers (called micro CMM or nano CMM) have been developed [1], such as the Molecular Measuring Machine developed by the National Institute of Standards and Technology (NIST) [2], the High-Precision Micro-CMM developed by the University of Tokyo and the National Institute of Advanced Industrial Science and Technology (AIST) [3], the special CMM developed by the Physikalisch-Technische Bundesanstalt (PTB) [4], the small-sized CMM developed by the National Physical Laboratory (NPL) [5], the Nanopositioning and Nanomeasuring Machine (NPMM) developed by the Ilmenau University of Technology [6–8], *etc.* The measurement range of these micro CMMs is not larger than 50mm in the X, Y and Z directions. Some other CMMs with a large measurement range have also been used in practice, for example, Zeiss F25 [9], Isara 400 [10], *etc.*

In this paper, an innovative micro CMM with zero Abbe error has been developed which includes some new design ideas, such as the self-made probe system, the metrological system and the co-planar stage. It has achieved the measuring uncertainty with 40 mm of about 100 nm (2σ).

2. Basic Structure and Key Technologies

2.1. Basic Structure

The schematic structure of the developed micro CMM is shown in Figure 1a. The measuring probe (1) is located under a granite column which is fixed on the granite base (5). The tip-ball of the stylus probe is placed at the center of the XYZ three-dimensional (3D) stage (2) and kept still after assembly. The sample is mounted on a moving table of the 3D stage and can be moved in X, Y and Z directions together with the stage. Two deadweight balance systems (4) are connected to the 3D stage and the granite base. They can transfer the deadweight of the 3D stage to the granite base. The position of each axis is measured by the corresponding laser interferometer (3). The reflection mirrors (7) and (8) of X- and Y-interferometers are fixed on two of the lateral sides of Y-stage. The reflection mirror of Z-interferometer is mounted on the bottom of Z-stage. The Z-interferometer and its reflection mirror are concealed and invisible in Figure 1. Figure 1b shows the photography of the main mechanical assembly of the developed micro CMM which is located in a thermostatic chamber isolated on an anti-vibration base.

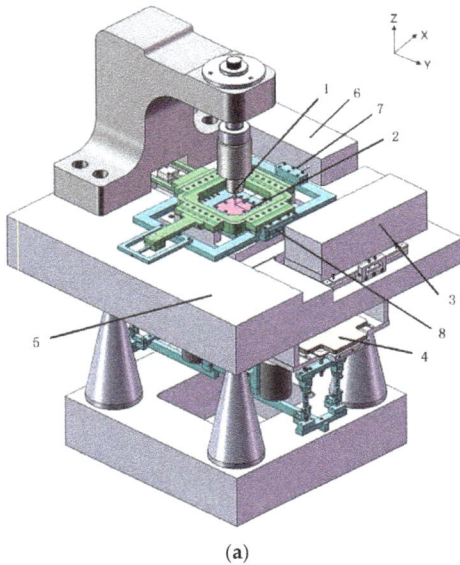

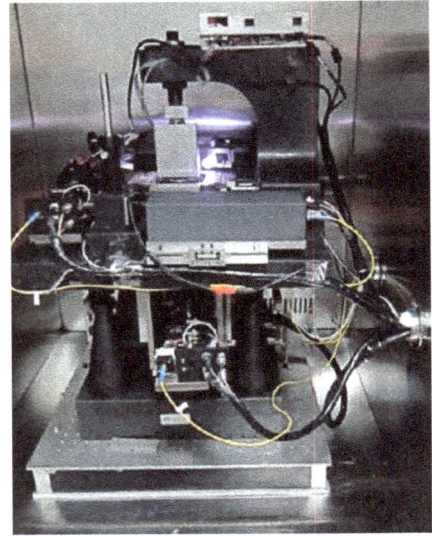

Figure 1. (**a**) schematic structure of the micro CMM (Coordinating Measuring Machines). (**b**) the photography of the mechanical assembly of the developed micro CMM. 1: Measuring probe, 2: XYZ stage, 3: Y-interferometer, 4: Deadweight balance, 5: Granite base, 6: X-interferometer, 7: Reflection mirror of X-interferometer, 8: Reflection mirror of Y-interferometer.

Abbe principle is the basis for all linear instrument designs. It is easy to obey the Abbe principle in a one-dimensional measurement but difficult to satisfy in all directions for a multi-dimensional measurement system. In conventional CMMs, measuring scales are always mounted on one side of each stage so that their metrological systems do not follow the Abbe principle. Abbe errors in conventional CMMs have to be reduced by well manufacture, fine adjustment in fabrication and error correction after assembly. However, limited by these efforts, the metrological systems of the conventional CMMs have difficulty achieving measurement accuracy better than 100 nm due to Abbe errors.

In order to follow the Abbe principle, a special structural design different from conventional CMMs is identified, as shown in Figure 2. The reflection mirrors (10) and (11) of X- and Y-interferometers are fixed on two of the vertical sides of the Y-stage (8). The reflection mirror of Z-interferometer (9) is mounted under the bottom of Z-stage (9). It can be seen that the length reference lines (represented by laser beams) of three axes (X, Y, Z) are perpendicular to each other and intersect at one fixed point, which is corresponding to the measuring point of the CMM. In addition, the X-guideway (5) and Y-guideway (7) are placed at the same height level and coincident with the reference plane (common plane) which is constructed by X-Y reference lines. All the guideways use roller bearing. The X-stage (6) and Y-stage (8) constitute a co-planar structure. These key components will be described in detail as follows.

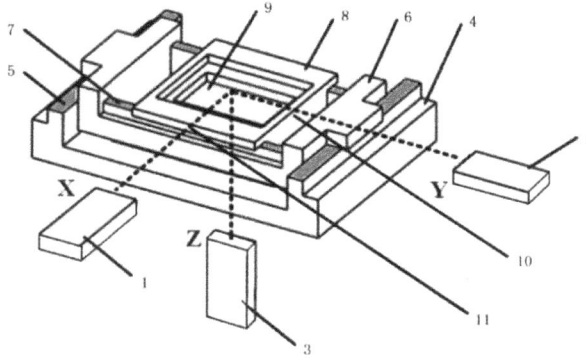

Figure 2. Schematic structure of 3D stage and schematic layout of metrological system. 1: X-interferometer (length reference), 2: Y-interferometer (length reference), 3: Z-interferometer (length reference), 4: Granite base, 5: X-guideway between X-stage and granite base, 6: X-stage, 7: Y-guideway between X- and Y-stages, 8: Y-stage, 9: Z-stage, 10: Reflection mirror of Y-interferometer, 11: Reflection mirror of X-interferometer.

2.2. Special Structure of the 3D Stage

Multi-dimensional stages can be constructed easily by stacking up several one-dimensional stages in series. However, the motion error is serious in this type of structure because the pitch angular errors of the bottom stage are magnified by the vertical offset between the guiding plane and the measurement point [11]. In order to reduce the motion error and the crosstalk between X- and Y-stages, a special mechanical structure is introduced as shown in Figure 2, where the X and Y bearings are almost co-planar and at the same height as the interferometer beam.

In Figure 2, the Z-stage (9) is embedded in the Y-stage and can be moved by roller bearing in the vertical direction. The Y-stage, together with the Z-stage, can be moved in the Y direction along the Y-guideway. The X-stage, together with the Y- and the Z-stages, can be moved in the X direction along X-guideway. The X-guideway between the granite base (4) and the X-stage shares the same horizontal plane with the Y-guideway, forming a co-planar stage. The Z-stage can be moved up and down through the X-Y co-planar stage. In this arrangement of the 3D stage, Abbe errors as well as crosstalk moving errors among X-, Y- and Z-stages are minimized. The strokes of the 3D stage in X, Y and Z directions are all 50 mm. The main body of the 3D stage is made of invar steel so as to reduce the thermal deformation of mechanical parts. The X-, Y-, Z-stages are driven respectively by three custom-made linear piezo-motors based on the type of N-310 provided by Physik Instrumente (Karlsruhe, Germany). The displacement resolution is about 1 nm.

2.3. The Metrological System and Its Layout

The metrological system consists of three laser heterodyne interferometers, developed by Zhang et al. [12,13] to measure the movements of the 3D stage in X, Y and Z directions, respectively. The reflectors of the interferometers are plane mirrors. The location of reflection mirrors are shown in Figures 1 and 2. The laser frequency of the three interferometers is stabilized at the level of about 10^{-7}. The resolution of the interferometer is 1 nm.

The layout of the metrological system is well arranged so that all interferometers are fixed on the granite base and independent from the 3D stage, as shown in Figure 1. The three measuring lines, represented by the three laser beams of the three corresponding interferometers, are parallel to the movement directions of the 3D stage and orthogonal to each other. They intersect at the measuring point, which is corresponding to the center of the probe's tip-ball, as shown by the dotted lines in Figure 2. The measuring lines of X- and Y-interferometers, X-guideway and Y-guideway are in the same plane. Therefore, such a metrological system obeys the Abbe principle and the crosstalk is minimized.

2.4. The 3D Measuring Probe

A contact probe [14] is adopted in this micro CMM. The physical and schematic structures of the probe are shown in Figure 3. The probe system mainly consists of an autocollimator (1), a mini Michelson interferometer (2), a suspension mechanism constructed by high sensitive elastic leaf springs (4), a reflection mirror (3) and a stylus with ruby ball (5). When the sample on the 3D stage contacts the probe ball and causes a deflection of the stylus, the reflection mirror will generate two angular displacements and a vertical displacement. Displacements of the mirror are functions of the linear deflection of the probe ball in X, Y and Z directions. The two angular displacements of the mirror are detected by the home-made autocollimator, and the vertical displacement is detected by the home-made Michelson interferometer. Then, the displacement of the probe ball in X, Y and Z directions can be calculated according to the output signals of the autocollimator and the interferometer. The measurement range of probe is up to 20 μm and the resolution is 1 nm in all X, Y and Z directions. The repeatability of the probe is better than 30 nm (2σ). By careful design and fabrication, the stiffness of the probe in three directions is nearly uniform. The maximum touch force is less than 12 mN when the probe ball is deviated from the original position by 20 μm. In other words, the stiffness is about 0.6 mN/μm.

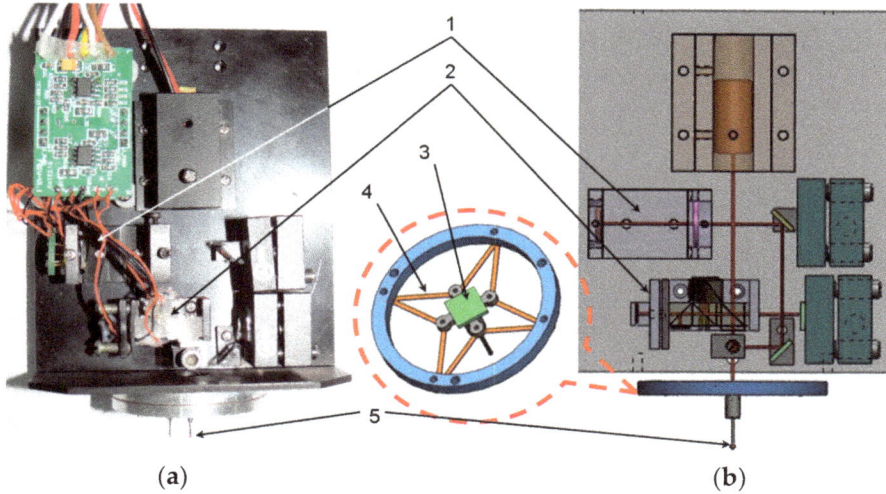

Figure 3. Physical and schematic structures of the 3D measuring probe. (**a**) physical picture of the probe; (**b**) schematic structure. 1: Autocollimator, 2: Michelson interferometer, 3: Mirror, 4: High sensitive elastic leaf spring, 5: Stylus with ruby ball.

2.5. Deadweight Balance Structure

The Z-stage is moved together with X- and Y-stages. The deadweight of the moving table will cause structure deflection in the XY-plane and uneven driving forces in the Z-axis due to gravity effect. In order to reduce this mechanical deformation and inertial force, two sets of counterweight mechanisms are proposed in the developed micro CMM. One is used to balance the deadweight of the Z-stage, which is called the Z-stage deadweight balance system. The other is to balance of the deadweight of X-, Y- and Z-stages during lateral movement, which is called the lateral deadweight balance system. Details are given as follows.

2.5.1. Z-Stage Deadweight Balance System

The Z-stage can move in the Z direction alone and in X and Y directions together with X-Stage and Y-stage, respectively. Its motion is driven by a commercial linear piezo-motor provided by Physik Instrumente. The deadweight of the Z-stage will induce more actuation force if it is moved up. In order to overcome this problem, a force balance system for Z-stage deadweight is particularly designed. Figure 4 shows the schematic structure. The Z-stage (1) and its supporting frame (5) are fixed together and driven by the linear piezo-motor (8). The Y-stage (2), pulley (3) and outer frame (6) are fixed together. Two counterweights (7) and the supporting frame (5) are connected by two thin steel strips (4). The total weight of two counterweights is equal to the net weight of the Z-stage and its accessories. In this mechanical structure, the Z-stage can be driven by the actuator with a low and equal driving force in both up and down motions.

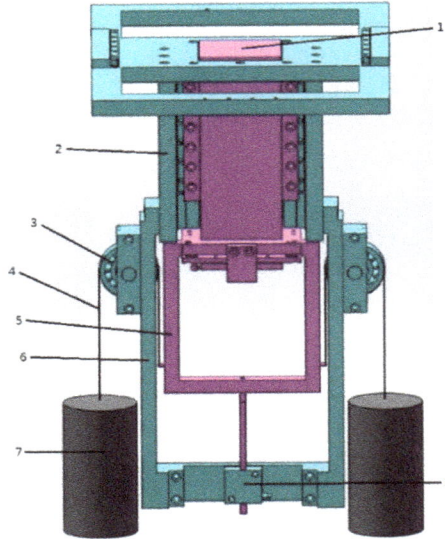

Figure 4. Schematic structure of Z-stage deadweight balance system. 1: Z-stage, 2: Y-stage 3: Pulley, 4: Steel strip, 5: Support frame, 6: Outer frame (It is fixed with Y stage), 7: Counterweight, 8: Linear piezo-motor.

2.5.2. Lateral Deadweight Balance System

In Figure 4, although the deadweight of the Z-stage is balanced by two counterweights, the total deadweight of the Z-stage with its accessories and its mechanical balance system is transferred to the Y-stage, then to the X-stage during Y and X motions. When the Y- or X-stage moves laterally, the total center of the gravity of the moving body moves and will cause structural deformation to the degree of micrometers, which is unacceptable in the developed machine. In order to balance the deadweight of the X- and Y-stages and the balance system of the Z-stage, we developed a lateral deadweight balance mechanism, as schematically shown in Figure 5. Through this mechanism, their deadweight is transferred to the granite base, which has a high stiffness.

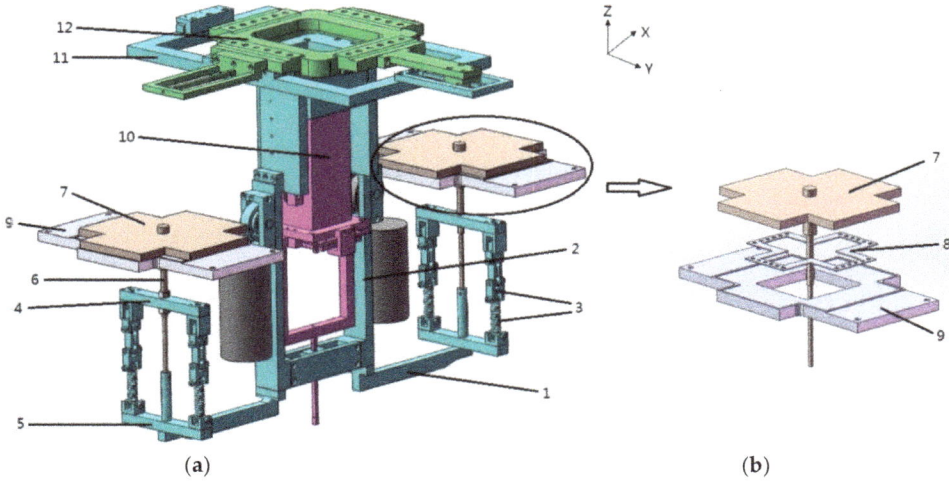

Figure 5. Schematic structure of the lateral deadweight balance system. 1: Horizontal arm, 2: Outer frame, 3: Spring and tension sensor, 4: Upper arm, 5: Lower arm, 6: Guide rod, 7: Upper plate, 8: Rolling bearing group, 9: Lower plate, 10: Z-stage, 11: Y-stage, 12: X-stage.

In Figure 5, the X-stage (12) is located on the granite base (not shown in this figure for clarity) and can move only along the X direction. The Y-stage (11) together with the balanced Z-stage can move not only in the Y direction along the Y-guideway but also in the X direction together with the X-stage.

Two horizontal arms (1) are fixed to the bottom of the outer frame (2) symmetrically and move together with the Y-stage. The lower arm (5), upper arm (4), springs and tension sensors (3), guide rod (6) and upper plate (7) are connected rigidly in the lateral direction. Thus, the movement of X- and Y-stages in the lateral direction is transferred to the upper plate (7). The upper plate, guide rod and upper arm are connected rigidly, and are also connected with the lower arm through two springs and tension sensors in each side. The tension forces of four springs are measured by four tension sensors respectively. During fabrication of the stages, the tension forces of the four springs are adjusted to be equal. In Figure 5b, the two-dimensional moving stage is composed of the upper plate, rolling bearing (8) and lower plate (9). The lower plate in each side is fixed rigidly to the granite base (not shown in this figure for clarity). The upper plates can move on the lower plates with a very low friction through the linear rolling bearing. Therefore, the total deadweight of all the moving parts is transferred to the granite base, and the stages can move in lateral direction smoothly.

3. Analysis and Correction of Main Errors

Although the developed micro CMM obeys Abbe principle in X, Y and Z directions, and some unique key technologies mentioned above are adopted to reduce systematic errors, some other errors are still obvious compared to the required measurement accuracy. These errors are frequency stability of the laser interferometer, flatness errors of long reflection mirrors for laser interferometers and non-orthogonal errors between interferometers. These errors are analyzed, separated and corrected in the following.

3.1. Stability of Frequency Stabilized Nd:YAG Laser Feedback Interferometer

In this micro CMM, movements in X, Y and Z directions are measured by three Nd:YAG laser feedback interferometers developed by Zhang *et al.* [12,13]. Frequencies of all three lasers are stabilized. The frequency stability error of each laser is less than 10^{-7}. Figure 6 shows the fluctuation of one interferometer' wavelength during 3–4 h. It can be estimated that the maximum measuring error caused by the frequency stability error is about 26 nm (2σ).

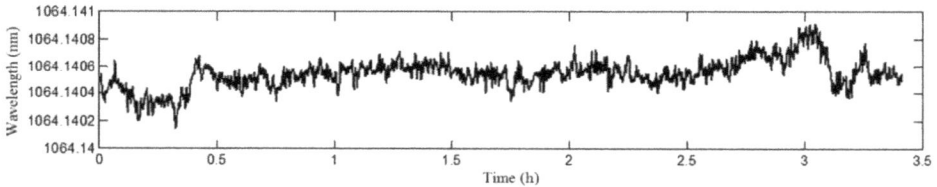

Figure 6. Calibration of the wavelength of one interferometer.

3.2. Error Calibration and Correction of Interferometers' Reflection Mirrors

The reflection mirrors of the three interferometers are fixed on the moving table. However, the reflection surfaces of mirrors themselves are not ideal flat and the surface deformation of each mirror is inevitable after the mirrors are fixed on the moving table. The straightness in X- and Y-mirrors and the flatness in Z-mirror will contribute to the final measuring result directly. Therefore, the errors of reflection mirrors should be separated and corrected. Figure 7 shows the schematic diagram of the straightness error separation of reflection mirror of X-interferometer. After the error calibration and correction, the measurement straightness of the optical flat is about 57 nm.

88

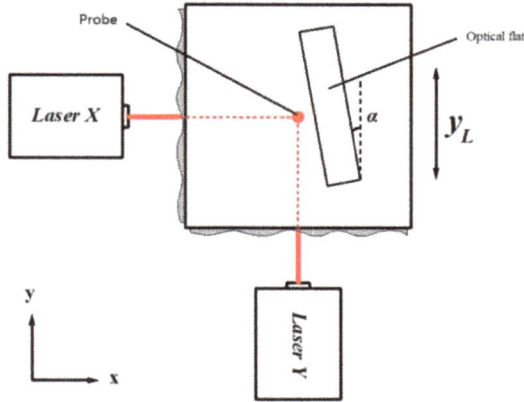

Figure 7. Schematic diagram of the surface flatness calibration of X-interferometer's reflection mirror.

In Figure 7, an optical flat with the accuracy of grade one is placed on the stage. The maximum flatness error of the optical flat's measuring surface is less than 30 nm. Here, the measuring surface of the optical flat is used as the reference and is adjusted parallel to the long reflection mirror surface to the best we can. In other words, the angle of α should be as small as possible. Then, the measuring surface of the optical flat is measured by the micro CMM. Using the obtained data, the straightness error of the X-interferometer's reflection mirror can be calibrated and its systematic error can be corrected. By this approach, the straightness of the Y-mirror and flatness of the Z-mirror are calibrated and corrected. After the error calibration and correction, all residual errors are less than 50 nm (2σ).

3.3. Separation and Correction of Non-Orthogonal Errors

According to the layout of our developed micro CMM, the surfaces of three reflection mirrors should be exactly orthogonal to each other. If it is not, non-orthogonal errors occur and increase the final measured error directly. Firstly, the guideways of the X-, Y- and Z-stages are adjusted to be orthogonal exactly by a calibrated square gauge. Then the interferometers are adjusted to keep the laser beam parallel with the X-, Y- and Z-guideways respectively. After that, the reflection mirrors are adjusted to be orthogonal according to the guideways. For example, Figure 8 shows the schematic diagram of the orthogonal error between reflection mirrors of X- and Y-interferometers. Provided that θ_{xy} is the non-orthogonal angular error between X- and Y-mirrors and the stage is moved by a displacement of x_0 in the X direction, the Y-interferometer will read an output error of Δy ($\Delta y = x_0 \cdot \tan\theta_{xy}$). This error will add to the Y coordinate of the measured point. In order to separate the non-orthogonal errors among three mirrors, the stage is moved in one direction only. Then, the outputs of other two interferometers are the combination of the flatness errors and non-orthogonal errors. Since systematic flatness errors of the mirrors are known because they have been corrected in the above section, then the non-orthogonal errors are obtained from the interferometers' outputs after the systematic flatness errors are subtracted. All the other non-orthogonal errors can be separated and corrected according to the same method. After the calibration and correction, all of the non-orthogonal errors are less than about 40 nm (2σ).

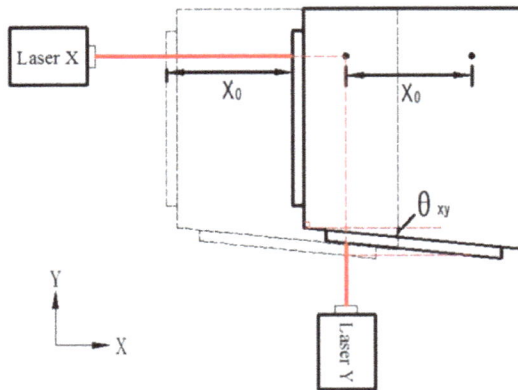

Figure 8. Schematic diagram of X-Y non-orthogonal error.

Besides the above errors, there are other remained errors. One is the thermal expansion error. The temperature control accuracy in the measuring area is $\pm0.5\,°C$. If the material of the ample (gauge block) is 12×10^{-6} and its length is 50 mm, then the thermal expansion error is about 30 nm (2σ).

The second is the residual Abbe error. The diameter of the probe ball used here is 1 mm. In Section 2.3, the center of the probe's tip-ball can not be adjusted at the right center of the laser beams of the three interferometers. This offset is within ±0.5 mm. These two offsets are Abbe offset. The motion angular error of X-, Y-stages is within $\pm6''$. Then, the Abbe error caused by this two offsets is about 30 nm (2σ). In Z direction, the maximum Abbe offset is the maximum measuring length 6 mm of the probe stylus. The motion angular error of X-, Y-stages is within $\pm2.1''$. Then, the Abbe error in Z direction is about 61 nm (2σ).

4. Performance Tests

In order to verify the measuring accuracy of the developed micro CMM, some high precision microparts, formed by gauge blocks with high grade, are adopted as samples to be measured. During the tests, the temperature of the thermostatic chamber is controlled at 20.00 °C ± 0.1 °C. Near the probe, the temperature is at 20.00 °C ± 0.05 °C. All of the gauge blocks used in this paper were calibrated with the accuracy of class 1 by the National Institute of Metrology, Beijing, China. The calibration uncertainty (2.96σ) of the gauge blocks is $(0.020 + 0.1 \times 10^{-6} \cdot l_n)$ µm. Here, l_n is the nominal length of the gauge block. Performance tests are introduced in the following.

4.1. Step Height Measurement in Z Direction

The step height in the Z direction was formed by the difference of two class 1 gauge blocks having calibrated lengths of 2.000022 mm and 5.000024 mm respectively, as shown in Figure 9. They were located together on the surface of the optical flat (the flatness error is less than 30 nm), which is fixed on the 3D stage. Six points on one surface of the gauge block (3) were measured by the developed micro CMM. A reference plane of it was then calculated by the least squares method. For the gauge block (2), the coordinate of the center point of the upper surface was measured. Then, the step height of the two gauge blocks is the distance between the center point and the reference plane. According to this method, the step height of the two gauge blocks was measured 15 times. The average value of the 15 repeated measurements was 3.000006 nm. The standard deviation (σ) was 21 nm. The average value 3.000006 mm of 15 repeated measurements is very close to the difference 3.000002 mm between the two calibrated values. During the measuring process, the triggering direction of the probe was in the vertical direction, so there was no need to compensate the probe radius.

Figure 9. Step height measurement schematic diagram between two gauge blocks. 1: Optical flat; 2: Calibrated gauge block 1; 3: Calibrated gauge 2.

91

The step height measurement errors are mainly from the interferometer (26 nm, $k = 2$), the residual Abbe error (30 nm, $k = 2$), the repeatability error of the probe (30nm, $k = 2$), the error from reflection mirror (50 nm, $k = 2$). Neglecting other errors sources, the measurement uncertainty is about 71 nm ($k = -2$).

4.2. Length Measurement in Lateral Direction

Because of the diameter error of the ball itself, the roundness deviation of the ball and the deflection of the ball stylus at triggering point, the equivalent diameter of the probe ball in the measuring direction should be known in advance. Here, the equivalent diameter of the probe in the measuring direction is calibrated firstly in the measurement direction by a reference gauge block with high accuracy. Then, the calibrated value is subtracted from the obtained data of the sample.

In the estimation of the equivalent diameter of the probe ball, six points on one measuring face of the reference gauge block (its calibrated value is 10.000045 mm) were measured by the developed micro CMM firstly. Using the coordinates of the six points, a reference plane was calculated by the least squares method. Then, one point on the opposite measuring face of the reference gauge block was measured. The distance between this point and the reference plane is the sum of the dimension of the gauge block and the equivalent diameter of the probe ball. Figure 10 shows the results of a 15 times repeated measurement (The horizontal coordinate is the order of measurement). Neglecting the dimensional error of the reference gauge block, the equivalent diameter of the probe ball in the X direction was calculated as $r_x = 11.001268$ mm $- 10.000045$ mm $= 1.001223$ mm. The standard deviation (σ) of the repeatability is about 26 nm.

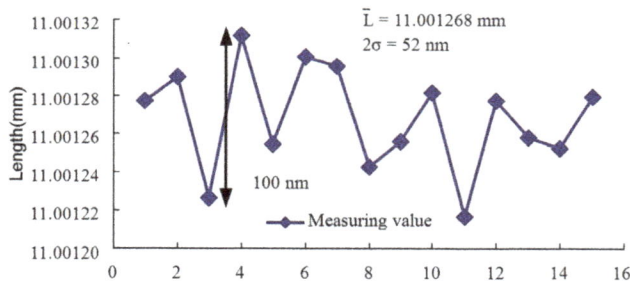

Figure 10. The repeated measuring results of sample gauge block.

After the equivalent diameter of the probe in the measurement direction had been calibrated, the reference gauge block was replaced by the sample gauge block in the same location and same direction. After the equivalent diameter of the probe ball was subtracted from the measured value, the measured dimension of the

92

sample gauge block could be obtained. Figure 11 shows 10 times measurement results of a sample gauge block with calibrated length of 40.000039 mm. The measured dimension is about 40.000011 mm on average and the standard deviation is about 47 nm.

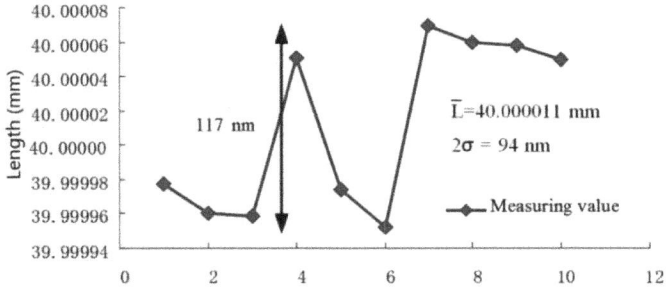

Figure 11. The measuring length of a gauge block with the nominal length of 40 mm.

Another test was done in the Y direction with the same setup as in the X direction. Figure 12 shows the probe diameter calibration results. The 10 times average equivalent diameter of the probe in Y direction is 0.998294 mm. The standard deviation (σ) of the repeatability is about 47 nm. Figure 13 shows 12 times measurement results of the same sample gauge block. The measured dimension is about 40.000054 mm on average and the standard deviation is about 45 nm.

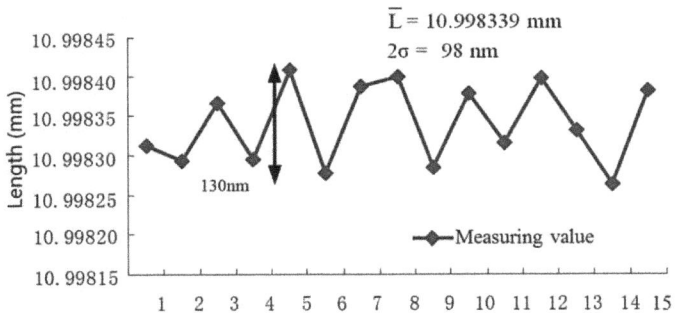

Figure 12. Calibration of the diameter of the probe ball by reference.

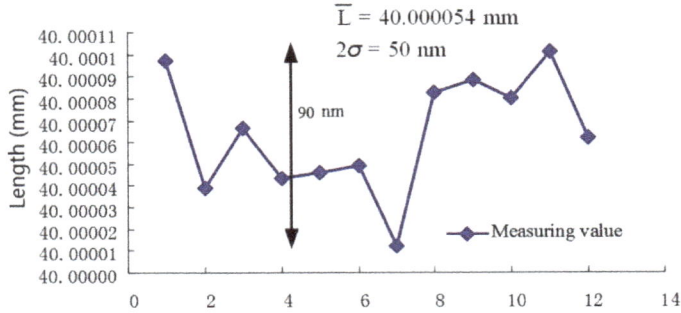

Figure 13. The measuring results of a sample gauge block.

In the lateral length measurement, the measurement errors are mainly from the interferometer (26 nm, $k = 2$), the thermal expansion error (24 nm, $k = 2$), the Abbe error (30 nm, $k = 2$) in lateral direction, the Abbe error in Z direction (61 nm, $k = 2$), the repeatability error of the probe (30 nm, $k = 2$), the error from reflection mirror (50 nm, $k = 2$), the non-orthogonal error (40 nm, $k = 2$). Neglecting other error sources, the measurement uncertainty is about 106 nm ($k = 2$).

Although the equivalent diameters of the probe ball in different directions shown in Figures 10 and 12 are different, but the measured dimensions of the sample gauge block, shown in Figures 11 and 13 have no obvious difference (about 43 nm).

4.3. Flatness Error Measurement

For verifying the flatness error of the developed micro CMM, a standard gauge block with the measuring face flatness of 30 nm was used to be measured. The gauge block was set on the Z stage directly. The probing direction is vertical. Six points are sampled on the surface of the gauge block with the measurement zone of 13 mm × 4 mm. The calculation of the flatness is based on the least squares method. The obtained coordinates of the six points are given in Table 1. The result of the flatness error was 66 nm. Other measuring surfaces of the gauge blocks were measured. Their flatness is in the same level.

Table 1. Obtained coordinates of 6 points on the measuring surface of a gauge block and flatness.

Point	Coordinates (x_i, y_i, z_i) (mm)	Point	Coordinates (x_i, y_i, z_i) (mm)
1	(18.004414, 34.999884, 34.741144)	2	(18.003694, 28.000295, 34.741274)
3	(18.002475, 22.000374, 34.741458)	4	(22.004273, 21.995507, 34.731724)
5	(22.005775, 27.999635, 34.731491)	6	(22.006136, 34.999752, 34.731304)
Flatness		66 nm	

In fact, we measured the flatness of an optical flat of Grade 0 with nominal diameter of 45 mm, which is the highest grade. We measured two areas of the measuring surface of the optical flat. One is 12 mm × 12 mm and the other is 28 mm × 28 mm. The obtained flatness errors were 56 nm and 62 nm, respectively.

5. Conclusions

This paper introduces a developed micro CMM with some new design ideas, especially with the metrological layout of being Abbe error free. In this machine, a structure of a co-planar stage is adopted so as to reduce the Abbe error and crosstalk among X-, Y- and Z-stages. A previously developed 3D measuring probe with the measuring range of 20 μm was adopted for use. The probe ball, the metrological system and 3D stage are carefully arranged so that the Abbe error is free. Experimental results show that the measurement uncertainty in X, Y and Z directions with 40 mm is about 100 nm (2σ) and the flatness error of measuring surface of the sample gauge block is about 66 nm.

This micro-CMM can only measure simple dimensions. For more complicated dimensions, further research is needed.

Acknowledgments: The authors acknowledge gratefully the support of the National High-Tech R & D Program (863 Program) (Project code: 2008AA042409) and the National Natural Science Foundation of China (Project code: 51475131 and 51175141).

Author Contributions: Qiangxian Huang established the layout of the micro system, and took part in the mechanical design, control system and experiments. Kui Wu contributed mainly to the control system and experiments. Chenchen Wang established the mechanical system and software, and took part in the experiments. Ruijun Li contributed to the probe system and took part in the experiments. Kuang-Chao Fan contributed to the probe system and co-planar stage. Yetai Fei took part in the establishment of the metrological system and mechanical system.

Conflicts of Interest: The authors declare no conflict of interest.

References

1. Claverley, J.D.; Leach, R.K. A review of the existing performance verification infrastructure for micro-CMMs. *Prec. Eng.* **2015**, *39*, 1–15.
2. Kramar, J.A. Nanometre resolution metrology with the Molecular Measuring Machine. *Meas. Sci. Tech.* **2005**, *16*, 2121–2128.
3. Yang, P.; Takamura, T.; Takahashi, S.; Takamasu, K.; Sato, O.; Osawa, S.; Takatsuji, T. Development of high-precision micro-coordinate measuring machine: Multi-probe measurement system for measuring yaw and straightness motion error of *XY* linear stage. *Prec. Eng.* **2011**, *35*, 424–430.
4. Brand, U.; Kleine-Besten, T.; Schwenke, H. Development of a special CMM for dimensional metrology on microsystem components. In Proceedings of the Sixth International Symposium on Precision Mechanical Measurements, GuiYang, China, 8 August 2013; pp. 542–546.
5. Lewis, A.; Oldfield, S.; Peggs, G.N. The NPL Small CMM-3-D measurement of small features. *5th LAMDAMAP.* **2001**, *34*, 197–207.
6. Manske, E.; Jäger, G.; Hausotte, T.; Füßl, R. Recent developments and challenges of nanopositioning and nanomeasuring technology. *Meas. Sci. Tech.* **2012**, *23*, 74001–74010.
7. Balzer, F.G.; Hausotte, T.; Dorozhovets, N.; Manske, E.; Jäger, G. Tactile 3D microprobe system with exchangeable styli. *Meas. Sci. Tech.* **2011**, *22*.
8. Schmidt, I.; Hausotte, T.; Gerhardt, U.; Manske, E.; Jäger, G. Investigations and calculations into decreasing the uncertainty of a nanopositioning and nanomeasuring machine (NPM-Machine). *Meas. Sci. Tech.* **2007**, *18*, 482–486.
9. Kornel, F.E.; David, B.; Martin, L.C.; Thom, J.H.; Thomas, R.K.; Marc, M.; Kamlakar, R.; Richard, D. *Micromanufacturing*; Springer: Dordrecht, The Netherlands, 2007; pp. 89–109.
10. Spaan, H.A.M.; Widdershoven, I. Isara 400 Ultra-precision CMM. In Proceedings of the Optical Fabrication, Testing, and Metrology IV, Marseille, France, September 2011.
11. Fan, K.C.; Fei, Y.T.; Yu, X.F.; Chen, Y.J.; Wang, W.L.; Chen, F.; Liu, Y.S. Development of a low-cost micro-CMM for 3D micro/nano measurements. *Meas. Sci. Tech.* **2006**, *17*, 524–532.
12. Tan, Y.D.; Zhang, S.L. Self-mixing interference effects of microchip Nd:YAG laser with a wave plate in the external cavity. *Appl. Opt.* **2007**, *46*, 6064–6048.
13. Tan, Y.D.; Zhang, S.L.; Zhang, S.; Zhang, Y.Q.; Liu, N. Response of microchip solid-state laser to external frequency-shifted feedback and its applications. *Sci. Rep.* **2013**, *3*.
14. Li, R.J.; Fan, K.C.; Miao, J.W.; Huang, Q.X.; Tao, S.; Gong, E.M. An analogue contact probe using a compact 3D optical sensor for micro/nano coordinate measuring machines. *Meas. Sci. Tech.* **2014**, *25*, 1–9.

Investigating Characteristics of the Static Tri-Switches Tactile Probing Structure for Micro-Coordinate Measuring Machine (CMM)

Yin Tung Albert Sun, Kuo-Yu Tseng and Dong-Yea Sheu

Abstract: This paper describes the fabrication of a series of micro ball-ended stylus tips by applying micro-EDM (Electrical Discharge Machining) and OPED (One Pulse Electrical Discharge) processes, followed by a manual assembly process of a static tri-switches tactile structure on a micro-CMM (Coordinate Measuring Machine). This paper further proves that the essential performance of the proposed system meets an acceptable benchmark among peer micro-CMM systems with a low cost. The system also adjusts for ambient temperature and humidity as the ordinary lab environmental conditions. For demonstration, several experiments used a randomly selected glass stylus with the diameters of stem and sphere of 0.07 mm and 0.12 mm, respectively. By leveraging research guidelines and common practice, this paper further investigates the probing relationship between measurement accuracy and its associated critical characteristics, namely triggering scenarios and geometric feature probing validation. The experimental results show that repeated detections in the uncertainty, in vertical and horizontal directions of the same point, achieved as small as 0.11 μm and 0.29 μm, respectively. This customized tri-switches tactile probing structure was also capable of measuring geometric features of micro-components, such as the inner profile and depth of a micro-hole. Finally, extensions of the proposed approach to pursue higher accuracy measurement are discussed.

Reprinted from *Appl. Sci.* Cite as: Sun, Y.T.A.; Tseng, K.-Y.; Sheu, D.-Y. Investigating Characteristics of the Static Tri-Switches Tactile Probing Structure for Micro-Coordinate Measuring Machine (CMM). *Appl. Sci.* **2016**, *6*, 202.

1. Introduction

Thanks to ongoing advancements in micro manufacturing technology over the past several decades, demand for micro-products, such as micro bio-medical and optical devices and MEMS (Microelectromechanical Systems) products, has been increasing significantly. To enhance manufacturing efficiency and improve quality for micro-products, high accuracy measuring devices are essential for micro fabrication technology. Conventional measuring devices, such as the Vernier calipers and micrometers, are incapable of measuring delicate micro-components. While a wide variety of optical measuring methods have been developed for measurement of micro-products, non-contact methods have yet to be developed for measuring the lateral-wall, high-aspect ratio and high reflection in micro-parts. More recently, in order to measure the geometry of micro-products, the micro coordinate measuring machine (μ-CMM) and a number of delicate tactile triggering structures have been developed [1–3]. However, the dimension of micro spherical stylus tips is one of the critical challenges for micro-CMMs. Although the micro tactile spherical styli with a diameter of 0.125 mm are commercially available, they are sold less often. To utilize a spherical stylus of such a small dimension, a micro-CMM with a high sensitive probing structure is necessary [3]. Due to the limitations of the stylus dimension, micro-CMMs still face some challenges when measuring the geometry of micro-components, such as micro-holes or micro-slots with a high aspect ratio. Micro ball-ended stylus tips, accomplished by a glue assembly process, held great precision with an eccentric deviation from 0.6 μm to 1 μm [4]. In our previous study, OPED (One Pulse Electrical Discharge) and gluing processes were successful to fabricate micro spherical stylus tips with a diameter of less than 0.1 mm [5]. To keep the strength of adherence sufficient for the micro-triggering touch, a Wire Electrical Discharge Grinding (WEDG) process was successful in producing micro stylus tips with diameters of less than 0.15 mm and the roundness of less than 1 μm, respectively, for the purpose of controlling the amount of glue applied between the glass ball and the stem [6].

The other critical problem with the micro-CMM is a sensing structure system. In the contact measuring, the stylus tip makes contact with the micro-object to trigger an electric signal. At the same time, the impact force is generated instantaneously. There is a correlation of size and fragility being the smaller the stylus, the more fragile it is. In our previous study, a micro stylus tip with a 40 μm stem diameter fractured when the impact force exceeded 11 mN [7]. To prevent them from breaking and fracturing easily during the measuring process, the sensing structure should have a sufficient stiffness in all direction [8]. Alblalaihid et al. applied 1 mN contact force to assess variable stiffness in the z-direction [9]. In addition, previous researchers reported various tactile sensing mechanisms of micro-CMM probing heads. However, these delicate triggering structures depended upon MEMS processes [10,11], which makes the probing heads so costly that they are not commercially available in the market. Our previous study reported a tri-switches tactile probing structure, focusing on only the vertical directional tactile triggering and few applications [12]. Subsequently, our other previous work enhanced both fabrication of micro-styli and multi-directional tactile triggering structure, facilitating the exploration of diverse measurement strategy [13]. Through system advancements, Alblalaihid et al. defined an associated measurement strategy for a specific probe [14,15] with stiff and flexible modes. It is observed that such an integral strategy is customized to a developed micro-CMM system while some benchmark parameters associated with the triggering scenarios are affected by the stiffness of a micro-probe, and are transferrable among micro-CMMs.

Conclusions drawn from the heretofore-mentioned systems provide segregated insight about the variable triggering scenario through a surface interaction to affect measurement accuracy, which is critical to the performance of micro-CMMs. Therefore, this paper not only refers to previously approved research guidelines and sound hands-on practice with a lean development in mind, but also investigates the probing relationship between measurement accuracy and critical characteristics such as triggering scenarios and geometric feature probing validation.

2. Materials and Methods: Probing Head Fabrication and Assembly

Because different measuring effects on the probed geometry stem from hardware construction, this section categorizes three aspects of stylus related fabrication.

2.1. Principle of Tri-Switches Probing Head Triggering Mechanism

To achieve a low-cost but high-sensitivity structure, the research team developed an enhanced sensing structure and tactile detecting mechanism from the previous work, as shown in Figure 1, to overcome the previously mentioned disadvantages and to achieve multi-directional measurement. Three micro wires of 50 μm in diameter were used to support the circle plate with 120 degrees apart. Three sensing rods of 150 μm in diameter were fixed upon the micro wires in each orthogonal direction. A micro spherical stylus tip was attached onto the circle plate using glue. The micro wires and rods maintain normal point-contacts in each direction when the stylus tip is in a standby condition. This means that the tri-switches of the micro stylus tip normally keeps the circuits closed without touching anything. However, when the stylus tip touches a micro-product, the three point-contacts switch to non-contact status instantly, and three detecting voltages change significantly. The micro-CMM system that was equipped with the proposed triggering mechanism of the tri-switches probing head, shown in Figure 2, used the electronic signal, to detect any reachable positions through triggering actions.

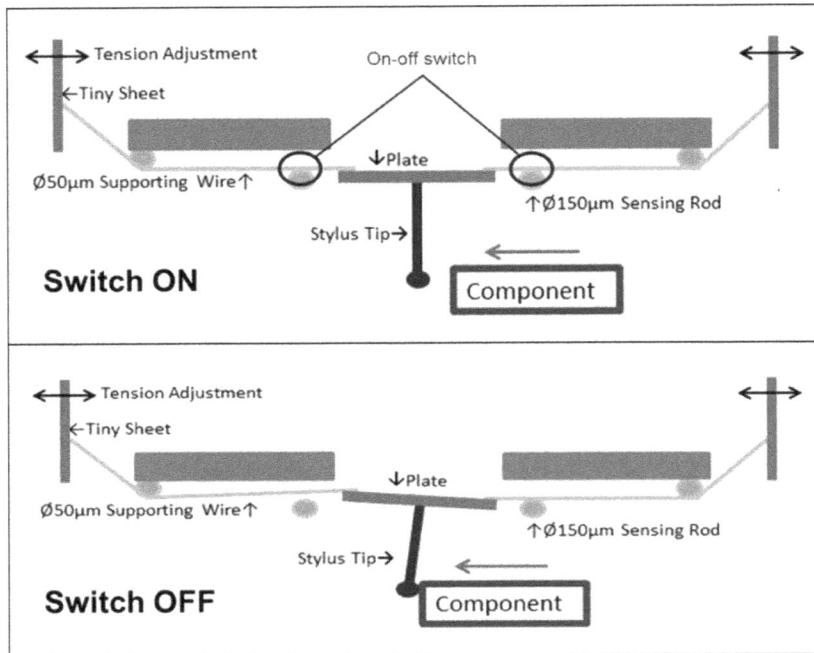

Figure 1. Tactile triggering mechanism of the proposed tri-switches probing head.

Figure 2. Tri-switches tactile probing structure.

2.2. Spherical Stylus Tips Fabrication

Micro spherical stylus tips play an important role in micro-CMM tactile probing heads. To detect the micro-component surface more precisely, the micro spherical stylus tips with a diameter less than 0.15 mm are necessary. In order to fabricate micro stylus tips, an assembly process called Wire Electrical Discharge Grinding (WEDG) and the OPED process were utilized. On the other hand, with properly setup parameters, the OPED process produced tungsten spherical stylus tips with consistently high precision. The scale of the aforementioned styli, as shown in Figure 3a,b, was suitable for the tri-switches probing head assembly detailed in the following section. In order to investigate the relationship between measurement accuracy and comprehensive characteristics, namely triggering scenarios and feature probing validation, a randomly selected spherical stylus was manually assembled onto the CMM probing head. In addition, the diameters of stem and sphere for the studied glass stylus were 0.07 mm and 0.12 mm, respectively, as shown in Figure 3b. The confirmatory results in Section 3 are applicable and representative for overall stylus tips mentioned in terms of the realistic measuring behavior of the constructed probing head in this study.

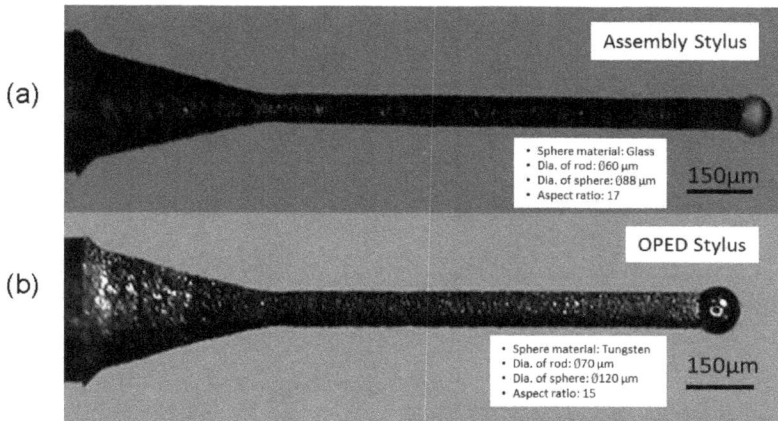

Figure 3. Stylus tips produced by an (**a**) assembly process called WEDG (Wire Electrical Discharge Grinding) process and the (**b**) OPED (One Pulse Electrical Discharge) process.

2.3. Probing Head Assembly Process

In order to assemble the micro spherical stylus tip onto the circle plate manually, two sets of x-y stages and one CCD (Charge Coupled Device) camera were used for position alignment, as shown in Figure 4.

Most commercial micro-CMMs are unable to use this tri-switches probing structure due to problems of compatibility and the extremely tiny stylus tip. To investigate measuring behaviors and evaluate the tri-switches probing structure, a three-axis high-resolution machine similar to the micro-CMM was set up, as shown in Figure 5a. This micro-CMM was controlled by a PC running a LabVIEW program (Version 11.0, National Instruments, Austin, TX, USA), as shown in Figure 5b. A laser interferometer was used to calibrate position accuracy of this similar micro-CMM;

the deviation of three-axis position accuracy was within the range of $\pm$ 3 µm. The tri-switches probing structure was mounted onto this micro-CMM to analyze its triggering behaviors. The micro-CMM, however, is clearly insufficient for high nano-order accuracy metrology. This research also focused on investigating the triggering possibility of the proposed probing head. The accuracy of this micro-CMM has improvement potential for high-accuracy measurement in the future.

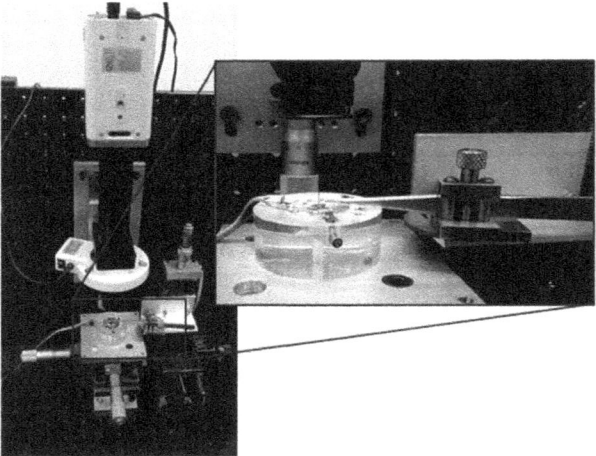

Figure 4. Manual assembly process for the tri-switches probing head.

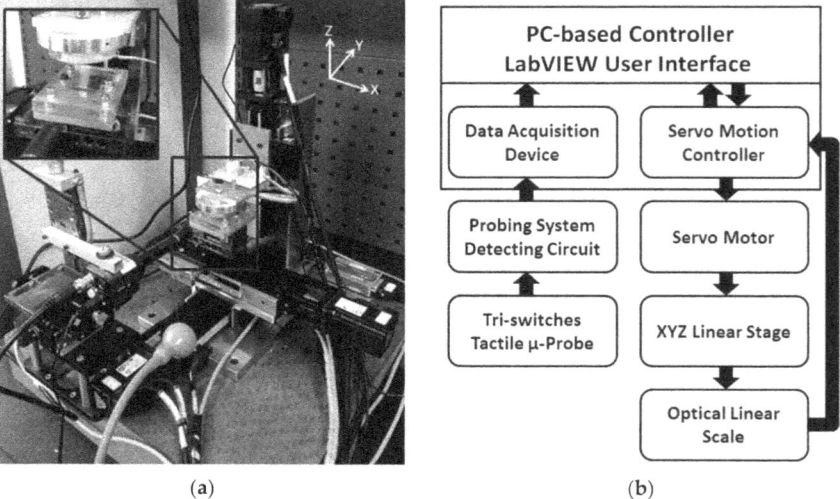

(a)   (b)

Figure 5. Scheme (**a**) and overview (**b**) of the developed micro-CMM (Coordinate Measuring Machine) system.

3 Results: Summarized Characteristics of Triggering Scenarios and Probed Feature Validation

The experiments involved two major investigations: (1) the triggering force, mechanism structure, and directional behaviors of the stylus; and (2) the measurement validation of the probed geometric features.

3.1. Triggering Force

In order to investigate the impact by a surface interaction, the triggering force of the tri-switches probing head was measured by a micro load cell which was mounted on the micro-CMM to measure the triggering force in the z-axis direction. After 30 contact repetitions, the average triggering force in the vertical direction was estimated as 1.27 ± 0.01 mN with 5% of risk, which met the stiffness requirement of the probe in all three directions and the isotropy of the probing force for triggering. Therefore, the micro stylus tip was able to endure the aforementioned impact force during the measuring process.

3.2. Mechanism Structure

The probing performance is dependent upon the mechanism structure, including the length of the micro stylus tip and the tension strength of the micro-wires. To measure micro-components with a high aspect ratio, the micro-CMM requires a longer stylus tip, as the sensitivity of tri-switches probing structure is determined by stylus tip length. In this study, two stylus tips consisting of a stainless steel plate and the electrical sensing module with the stem lengths of 5 and 10 mm were used to investigate triggering behaviors. While the performance of triggering behaviors on the z-axis with these two stylus tips are theoretically identical, triggering accuracy is compromised in the horizontal direction, as shown in Figure 6. To prevent the undesirable elastic deformation of the bending stylus impeding the triggering accuracy, the stylus of 5 mm was used. Although using a finite element model (FEM) to study the relationship between applied sensing load and probe stiffness is not within the scope of the study, a trend of using FEM facilities has led to a similar development approach [15].

3.3. Triggering Behaviors in Vertical and Lateral Directions

Following completion of the probing head assembly, the triggering characteristics of this probing structure on the same point were investigated through repeated detecting trials. For the vertical direction, as shown in Figure 7, the detection of the z-axis triggering position resulted in detection uncertainty to an appropriate extent. Therefore, tri-switches probing structure was capable of measuring geometry profile with micro-scale height difference in the vertical direction. Due to the cost associated with uncertainty test referring to both types A and B with reference to GUM (Guide to Uncertainty in Measurement): resolution of the instruments, temperature variation, variation in position detection, etc. was consolidated into data collection performed by common practice.

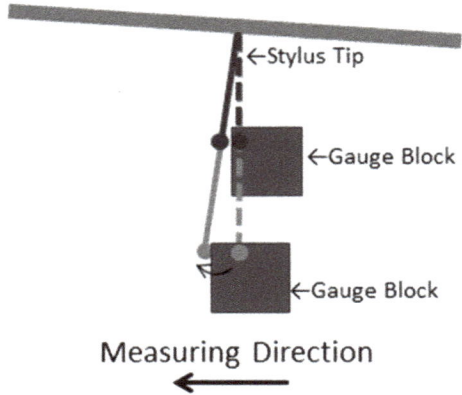

Figure 6. Avoiding the elastic deformation of stem bending which creates an undesirable triggering behavior.

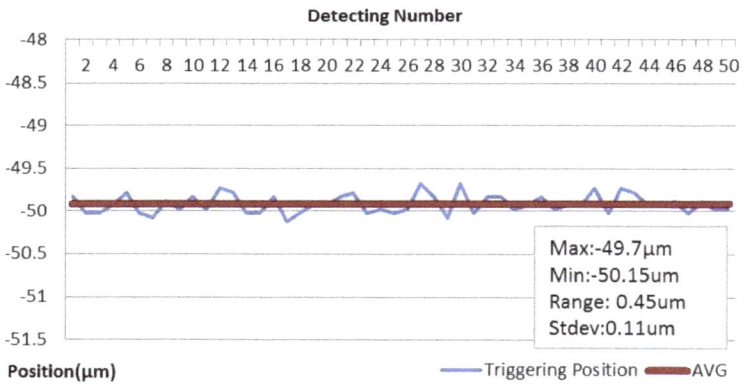

Figure 7. Detection uncertainty in the vertical z-axis direction.

As for the horizontal direction, the lateral triggering characteristics of this probing structure on the same point were investigated through repeated detecting trials in predetermined six angulated divisions, as illustrated in Figure 8. The results about the uncertainty appeared less than 1 µm in nine out of twelve angle graduations, shown in Figure 9. Although the deviation of the lateral triggering behaviors at 90, 120, and 150 degree graduations were relatively larger than the rest, the uncertainty enhancement compared to the former model [12] was evident.

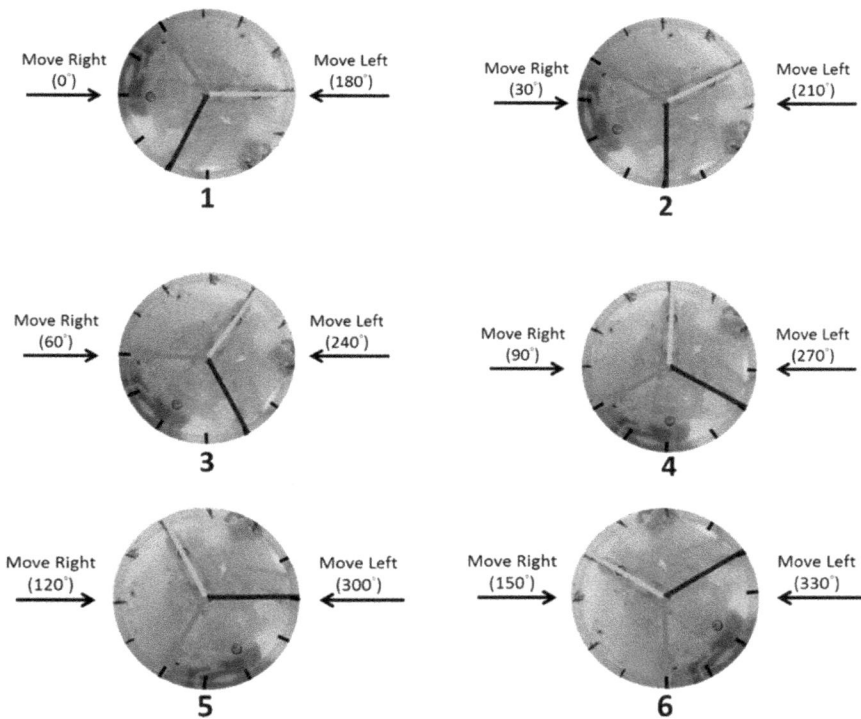

Figure 8. Lateral triggering measurements in six angulated divisions.

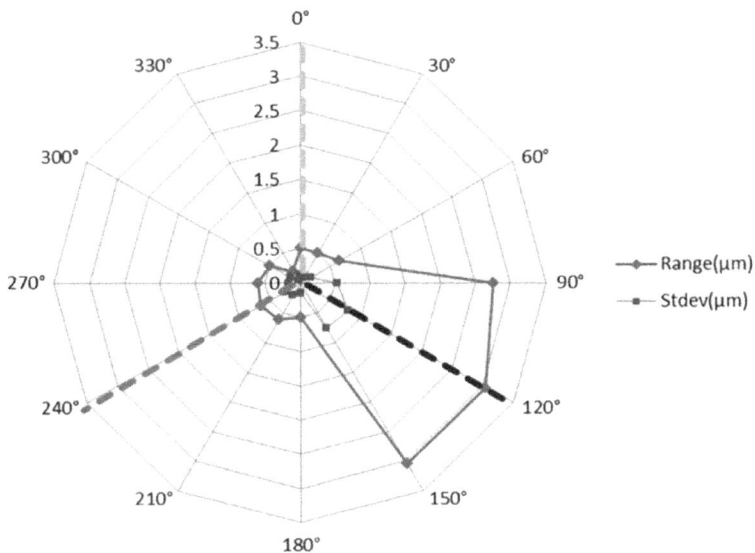

Figure 9. Lateral triggering detection uncertainty.

Such a behaving scenario most likely stemmed from the unbalanced tension strength of the supporting wires, which were incapable of recovering their initial positions. As an improvement, adding an auxiliary sensing wire structure may resolve the issue to achieve higher accuracy measurements.

3.4. Surface Interaction on Different Materials of Spheres

In dealing with probing head systems in the order of micrometers, the phenomenon of surface interaction needs to be considered. The effect of surface interaction shows rising significance and cannot be disregarded, especially when using a smaller-scale stylus tip to measure micro-products.

Two types of processes were applied to fabricate micro spherical stylus tips in this study. The materials of the sphere were glass and tungsten in the assembly process and the OPED process, respectively. Two stylus tips with sphere material were used to investigate measuring behaviors on different material within the component. The component is a gauge block made of metal and a spherical LED (Light-emitting Diode) lens made of resin, respectively. A sticking force can be observed when the micro sphere comes off of the feature surface during horizontal direction measurement. The experimental results showed that the glass sphere produced a significant oscillation that might impact a sensing wire structure causing measurement deviation, as shown in Figure 10. Although sticking force between metal sphere and spherical LED resin was observed to be extremely minute, with virtually no oscillation when measuring the gauge block with a metal sphere, the potential influence of the static electricity caused by insufficient isolation about remnant (static) charges between the gauge block and the sphere should be investigated in the future extended study. Because of the oscillation phenomenon and sufficient hardness, this study utilized the WEDG and OPED processes to fabricate glass and tungsten micro spherical stylus tips, respectively.

3.5. Geometric Profile Measurement

In this study, the tri-switches probing structure was used in several measuring examples. Measuring the step between two different gauge blocks illustrated one case. One of the critical problems in precision measurement was temperature, since maintaining a steady temperature in the environment tended to be difficult. This study used the same material gauge blocks that had similar thermal expansion coefficients in order to decrease deviation. These two gauge blocks were mounted on a granite platform, as shown in Figure 11. The results showed that the respective averages of the step were 252.45 μm and 253.83 μm in the vertical direction and 0 degree in a lateral direction. The deviations were caused by many factors, including machines, programs, calibration, environments, and probing heads.

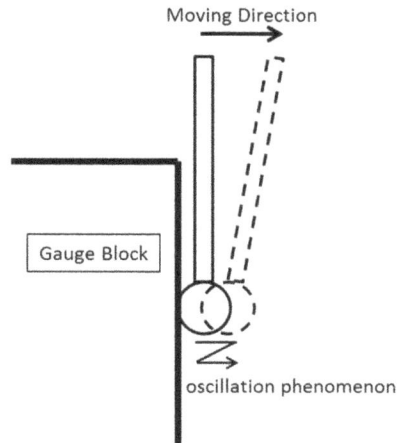

Figure 10. Oscillation phenomenon occurred while measuring.

Figure 11. Gauge blocks mounted on the granite platform.

3.6. Micro-Component Geometric Feature Measurement

Two kinds of micro-component geometric features, namely hole profile and depth, were studied to validate the performance of the micro-probing structure. Precision geometric measurement technologies have been under development for decades. In the case of micro-components with high aspect ratios, high optical-reflection and high curvature surfaces, however, micro-CMMs are still more reliable than an optical metrology device, particularly because these optical metrology devices are still incapable of precisely measuring the geometry profile of an inside micro-hole with a diameter of less than 0.2 mm. Nevertheless, these micro-holes with high aspect ratios are widely applied to engine nozzles, printer nozzles and wire drawing dies utilized in the textile industry. In this study, therefore, the inside geometry profile of the micro-hole was measured by the proposed tri-switches probing structure. The experiment drove the probe to collect hits across numerous depths in discrete movements with an upward increment of 5 μm. The profile detecting results were consistent with the inner geometry of the micro-hole according to the section observation of the optical metrology device, as shown in Figure 12.

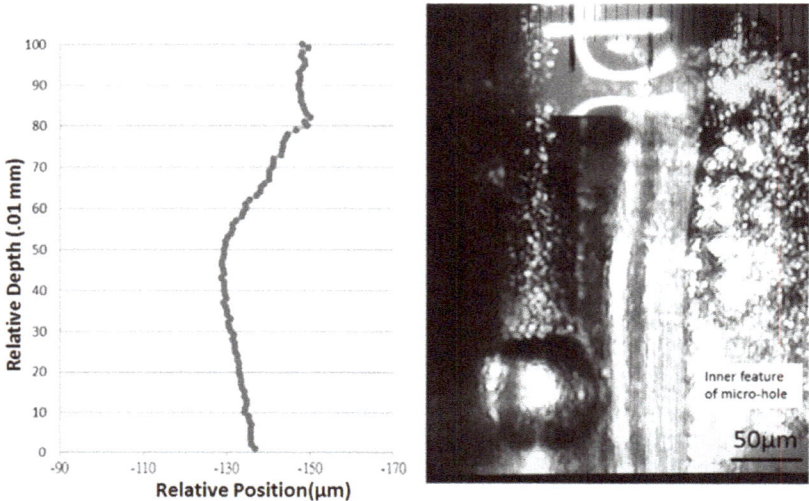

Figure 12. Consistency of measuring the inner profile of a micro-hole: the proposed vs. optical device.

The depth of micro-holes in this study was measured by the tri-switches probing head. The depth of micro-holes drilled by a Ø0.3 mm micro drilling tool could be easily measured by a spherical stylus tip with sphere diameter of Ø120 μm, as shown in Figure 13.

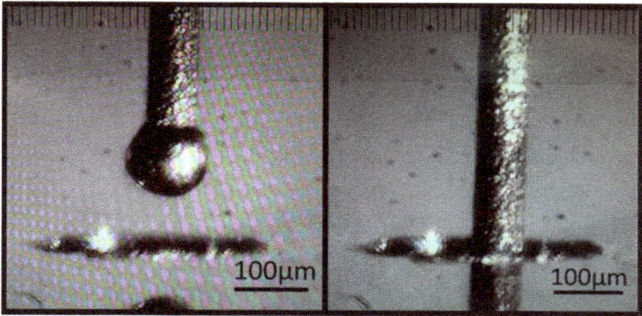

Figure 13. Depth of a micro-hole being measured by the probing head.

High-accuracy performance in the vertical direction of the tri-switches probing head was also validated with this experimental result, which involved 30 repeated measurements of the depth of a micro-hole. The average depth was estimated 222.82 ± 0.20 μm with 5% risk. However, compared to the results from using an optical microscope (the within uncertainty was close to 20 μm), as shown in Figure 14, the differential value was as much as 15.43 μm. It might be due to the lip clearance angle of the micro drilling tool. Because the inside bottom of the micro-hole was not leveled, measurement results by these two devices tended to be inconsistent. Otherwise, the difference between them could be more ideal. Therefore, the height feature that was measured by this micro-CMM system maintained more favorable reproducibility than the optical metrology device.

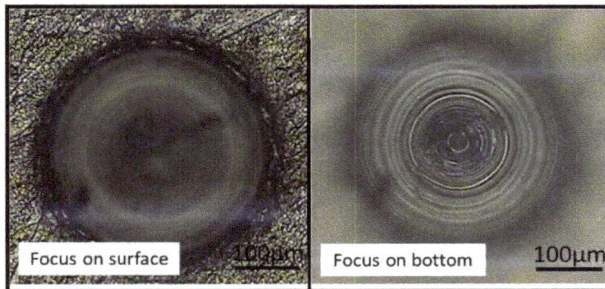

Figure 14. Measurement results by optical microscope.

4. Discussion

The technology introduced in this paper combines fabrication and assembling processes to develop a novel probing head that is mounted on a micro-CMM and is capable of multi-directional measurement on geometric features of a micro-part. As explained, the resulting sensing structure and characteristics by this tri-switches probing head structure indicate how it will be possible to measure micro-parts.

The tactile probing structure for micro tri-switches is composed of a distinctive sensing wire structure and an assembled micro spherical stylus tip, in which the diameters of the stem and sphere are 0.07 mm and 0.12 mm, respectively. In experiments on the same point with repeated detecting, repeatability was as low as 0.11 μm in the vertical direction. Repeatability deviation and triggering displacement in a lateral direction was less than 3 μm and 50 μm, respectively. In experiments measuring inside geometry profiles of micro-holes, the results proved that the developed tri-switches probing structure was practicable for measuring the surface features of micro-components. In lateral measurement, the effect of unbalanced tension strength from supporting wires and improper position of the sensing rod could possibly influence triggering behaviors. Therefore, to achieve higher accuracy measurement, future improvements may include an auxiliary sensing wire structure in order to take advantage of the low-cost static tri-switches triggering structure for measuring the geometry and profiles of micro-products. As for the observed sticking force, isolations from remnant static electricity, van der Waals force, air currents, temperature variation and external vibrations, to name a few, should be considered in future system evaluations. In addition, using FEM facilities to predict the mechanical deformation of all mechanisms under load is also part of the extended future research work.

Author Contributions: Dong-Yea Sheu and Yin Tung Albert Sun conceived and designed the experiments; Kuo-Yu Tseng performed the experiments; Yin Tung Albert Sun and Dong-Yea Sheu analyzed the data; Yin Tung Albert Sun and Kuo-Yu Tseng wrote the paper.

Conflicts of Interest: The authors declare no conflict of interest.

References

1. Küng, A.; Meli, F.; Thalmann, R. Ultraprecision micro-CMM using a low force 3D touch probe. *Meas. Sci. Technol.* **2007**, *18*, 319–327.
2. Claverley, J.D.; Leach, R.K. A vibrating micro-scale CMM probe for measuring high aspect ratio structures. *Microsyst. Technol.* **2010**, *16*, 1507–1512.
3. Bos, C. Aspects of tactile probing on the micro scale. *Precis. Eng.* **2011**, *35*, 228–240.
4. Hansen, H.N.; Carneiro, K.; Haitjema, H.; De Chiffre, L. Dimensional micro and nano metrology. *CIRP Ann.-Manuf. Technol.* **2006**, *55*, 721–743.
5. Cheng, C.C.; Huang, J.K.; Sheu, D.Y. Development of gluing and assembling process on micro EDM to fabricate micro spherical stylus tips for micro-CMM. In Proceedings of the 6th International Conference on MicroManufacturing ICOMM, Tokyo, Japan, 7–10 March 2011; pp. 219–222.
6. Sheu, D.Y.; Cheng, C.C. Assembling ball-ended styli for CMM's tactile probing heads on micro EDM. *Int. J. Adv. Manuf. Technol.* **2013**, *65*, 485–492.
7. Sheu, D.Y. Micro-spherical probes machining by EDM. *J. Micromech. Microeng.* **2004**, *15*, 185–189.
8. Fan, K.C.; Cheng, F.; Wang, H.Y.; Ye, J.K. The system and the mechatronics of a pagoda type micro-CMM. *Int. J. Nanomanuf.* **2012**, 67–86.
9. Alblalaihid, K.; Kinnell, P.; Lawes, S. Fabrication and characterisation of a novel smart suspension for micro-CMM probes. *Sens. Actuators A: Phys.* **2015**, *232*, 368–375.
10. Brand, U.; Kleine-Besten, T.; Schwenke, H. Development of a special CMM for dimensional metrology on microsystem components. In Proceedings of the 15th Annual Meeting of the American Society for Precision Engineering, Scotsdate, AZ, USA, 22–27 October 2000; Volume 542, pp. 1–5.
11. Claverley, J.D.; Sheu, D.Y.; Burisch, A.; Leach, R.K.; Raatz, A. Assembly of a novel MEMS-based 3D vibrating micro-scale co-ordinate measuring machine probe using desktop factory automation. In Proceedings of the 2011 IEEE International Symposium on Assembly and Manufacturing (ISAM), Tampere, Finland, 25–27 May 2011.

12. Kao, S.M.; Sheu, D.Y. Developing a novel tri-switch tactile probing structure and its measurement characteristics on micro-CMM. *Measurement* **2013**, *46*, 3019–3025.

13. Sheu, D.Y. Manufacturing tactile spherical stylus tips by combination process of micro electro chemical and one-pulse electro discharge technology. *Int. J. Adv. Manuf. Technol.* **2014**, *74*, 741–747.

14. Alblalaihid, K.; Kinnell, P.; Lawes, S.; Desgaches, D.; Leach, R. Performance assessment of a new variable stiffness probing system for micro-CMMs. *Sensors* **2016**, *16*, 492.

15. Alblalaihid, K.; Lawes, S.; Kinnell, P. Variable stiffness probing systems for micro-coordinate measuring machines. *Precis. Eng.* **2016**, *43*, 262–269.

A Micro-Coordinate Measurement Machine (CMM) for Large-Scale Dimensional Measurement of Micro-Slits

So Ito, Hirotaka Kikuchi, Yuanliu Chen, Yuki Shimizu, Wei Gao, Kazuhiko Takahashi, Toshihiko Kanayama, Kunmei Arakawa and Atsushi Hayashi

Abstract: This paper presents a micro-coordinate measuring machine (micro-CMM) for large-scale dimensional measurement of a micro-slit on a precision die coater by using a shear-mode micro-probe. A glass micro sphere with a nominal diameter of 52.3 μm was attached on one end of a tapered glass capillary tube as a probe tip ball. The micro-slit width of a slot die coater with a nominal slit width of 85 μm was measured by the micro-CMM. The probe tip was placed in the slit for the measurement. The effective working length of the probe was confirmed experimentally to be at least 1 mm. In order to measure the gap width uniformity over the entire slot die length of 200 mm, an air-bearing linear slide with a travelling stroke of 300 mm was employed in the micro-CMM to position the probe along the length direction of the slot die. The angular alignment error and the motion error of the air-bearing linear slide as well as those of the stages for positioning the probe along the direction perpendicular to the length direction of the slot die were investigated for evaluation of the expanded uncertainty of gap width measurement.

Reprinted from *Appl. Sci.* Cite as: Ito, S.; Kikuchi, H.; Chen, Y.; Shimizu, Y.; Gao, W.; Takahashi, K.; Kanayama, T.; Arakawa, K.; Hayashi, A. A Micro-Coordinate Measurement Machine (CMM) for Large-Scale Dimensional Measurement of Micro-Slits. *Appl. Sci.* **2016**, *6*, 156.

1. Introduction

Precisely machined micro-features have been utilized widely with the development of micro-fabrication technologies such as micro-cutting [1,2], micro-electrical discharge machining (EDM) [3,4], micro- and nano-lithography [5,6], nanoimprint [7] and micro-molding [8]. In particular, more precision micro-features with high aspect ratios represented by deep micro-holes and micro-slits are utilized for precision tools and measuring instruments that require sub-micrometric or nanometric dimensional accuracies. The requirements for dimensional measurements of precision micro-features with high aspect ratios are thus increasing. The micro-slit of a slot die coater is a good example of such precision micro-features. The slot die is a precision tool for coating of functional liquid materials on the surface of a substrate such as a glass plate or a film. On the top of the slot die coater is a micro-slit from which the liquid material is extruded. The micro-slit has a high aspect ratio with a gap width in the order of 100 μm and a depth of more than 10 mm. Since the gap width of the micro-slit as well as its uniformity along the length direction of the slot die coater is a key parameter for determining the coating quality, it is necessary to measure the micro-slit gap width and its uniformity accurately.

For dimensional measurement of micro-features with high-aspect ratios, the micro-coordinate measurement machine (micro-CMM) with a micro-probing system is one of the prospective measuring instruments because the inside of a micrometric feature such as a micro-hole or a micro-slit can be detected by the micro-probe with a micro-tip ball. On the other hand, the accuracy and reliability of the micro-probing system are determined by the positioning accuracy of the probe qualification of the probe tip, measurement strategies, and so on. The plastic deformation caused by the contact force between the probe and the measured surface is also a main error source, which requires the probing system to have a high sensitivity for detecting the probe–surface contact so that a small contact force can be achieved for reduction of the plastic deformation [9].

The contact detection is typically carried out by detecting the elastic deformation of the springs for supporting the stylus of a micro-probe. Different kinds of sensors have been integrated into the micro-probe for this purpose. For example, three integrated strain gauges have been employed to detect the probe–surface contact [10]. Optical systems have also been integrated into the micro-probes for contact detection. Cui *et al.* have proposed a three-dimensional fiber probe for the measurement of micro parts. [11]. Muralikrishnan *et al.* have proposed a fiber probe with an ellipsoidal tip for three-dimensional measurement on micro-features [12]. Fan *et al.* have developed low-cost and high precision micro-CMM by utilizing the focusing probe [13]. Schwenke *et al.* have developed an opto-tactile sensor for the contact detection of a micro-probe in two and three directions at Physikalisch-Technische Bundesanstalt (PTB) in Germany [14], where the micro-probe has a tip ball of 60 μm in diameter. The probe displacement can be detected by using an optical beam focused on the probe tip ball. Takaya *et al.* have proposed a nano-probe system based on the laser-trapping technique [15,16]. A micro-sphere with a diameter of 8 μm, which is held by the focused laser beam, is used as the probe tip ball with a piconewton measuring force. Such optical-based detection methods are more sensitive than the traditional strain gauges. However, the effective working distance of the probe is confined by the numerical aperture of the focusing lens and it is difficult to measure the inside of a high-aspect ratio micro-feature.

Detection of electric contact has been applied for micro-probing systems for highly sensitive probe-surface contact detection. Masuzawa *et al.* have introduced a micro-probe, in which the probe-surface contact is detected based on the electrical contact. Since the a micro-probe tip is attached on the end of a long and sharp stylus shaft, the probe can be used to measure the inside of micro-holes [17,18]. The micro-probe tip has a pyramidal edge and it can detect the contact in one direction. As for the multi-axis detection of contact, Weckenmann et al. have developed a probing technique based on the method of scanning tunneling microscopy (STM) [19], in which a spherical probe tip is employed. However, the applications of such methods are only effective for measurement of conductive materials.

Improvement of sensitivity for contact detection can also be achieved by applying mechanical vibrations to the micro-probe. Bauza et al. have proposed a vibrating shank for measurement of the inside of a micro-hole [20,21]. A fiber with a diameter of 7 μm is vibrated in one direction by a crystal resonator, which is used as a virtual micro-probe for contact detection in one axis. The effective diameter of the vibrating fiber probe is approximately 30 μm. With respect to a probe with multi-axis detection, Claverley et al. have developed a three-dimensional vibrating probe composed of PZT sensors and actuators [22,23]. The probe tip of the three-dimensional vibrating probe has a spherical tungsten tip with a diameter of approximately 70 μm. Murakami et al. have presented an optical fiber probe which has a glass tip ball with a diameter of 5–50 μm [24,25]. Since the vibrating probe allows the use of a probe stylus with a lower stiffness, it results in a lower probe–surface contact force. In addition, it is an effective means for the vibrating probe tip to reduce the influence of adhesion force of the measured surface against the probe. However, the measurement is influenced by the water layers on both the measured surface and the probe surface [9,22,25]. In responding to this problem, the authors have developed a shear-mode micro-probe, which can reduce the influence of the water layers through detecting the interaction force between the probe and the measured surface over the water layers [26]. It has been demonstrated that the shear-mode detection probe can achieve nanometric resolution and probing repeatability.

The motivation of this research is to apply the shear-mode micro-probe for measurement of the gap width of the micro-slit on a precision slot die coater. The precision slot die is composed of two precision flat plates [27] made of stainless steel with a length of 200 mm. The two precision flat plates are separated with a gap with a nominal gap width of 85 μm to form the micro-slit on the coater edge. It is required to make gap width measurement of the micro-slit with the expanded uncertainty of less than 100 nm [28]. Traditionally, mechanical feeler gauges have been used for the measurement of micro-slit gap width. However, the mechanical feeler gauge often induces damages on the edge areas of the micro-slit due to the large contact force. Although a CCD camera based optical system [28] has been proposed for non-contact measurement of the micro-slit gap width, it is difficult for the optical system to achieve sub-micrometer resolutions because of the diffraction limit of the

114

optical system. As for the detection of single edge position, it can be achieved by using the optical system. However, with regard to the precision measurement of the slit gap width, it is necessary to detect the positions of two edges quantitatively. Therefore, the resolution in lateral direction is one of the important requirement for the gap width measurement. In addition, it cannot be used for measuring the gap width inside the micro-slit. Recently, a non-contact measurement system with a higher resolution by using a capacitive displacement sensor has been developed [27]. However, the measurable minimum gap width of the electric gap sensor is limited to 150 μm, due to the sensor thickness. Compared with the measurement techniques mentioned above, a micro-CMM is more effective for such a measurement in terms of the measurable minimum gap width and measurement resolution, as well as the flexibility and accessibility to the inside of the micro-slit.

This paper presents a micro-CMM specifically designed for the slit width measurement of the precision die coater. In addition to the newly developed shear-mode micro-probe, an air-bearing linear slide with a travelling stroke of 300 mm was employed in the micro-CMM so that there is a uniformity of the gap width over the entire length of the micro-slit of 200 mm. This is a significant difference from an existing micro-CMM that can only cover a measurement length from several millimetres to several centimetres. The calibration of the conventional micro-CMMs is carried out by using the calibration artifact with a length of 20 mm at the maximum [29]. Therefore, a longer travelling range of the probe positioning system is essential for the gap width measurement of the slot die. The effective working length of the shear-mode micro-probe is designed to be 1 mm, which is longer than that of the existing micro-probes [25,30], so that the inside of the micro-slit can be accessed by the probe tip while avoiding the influence of the chamfered edges of the precision parts forming the micro-slit. In order to shorten the measurement time, a measurement strategy of gap width measurement based on two probing points is also introduced. Next, a description of the experimental setup for the micro-CMM, experiments and uncertainty analysis of the measurement results are presented.

2. Experimental Setup for the Micro-CMM

Figure 1a shows the photographs of the shear-mode micro-probe. The micro-stylus is composed of a glass micro-stylus shaft and a glass micro-sphere. The stylus shaft is made of a capillary glass tube which is thermally pulled by using a commercially available glass pipette puller. A micro-glass sphere with a nominal diameter of 52.6 ± 3.2 μm (Certified size standard Cat No. 9050, Duke Scientific Corporation, California, CA, USA) [31] is attached on one end of the glass micro-stylus by thermosetting resin. The diameter of the commercially available glass micro-sphere was selected based on the nominal gap width of the precision slot die. Figure 1b

115

shows the schematic of a shear-mode micro-probe. The micro-stylus is then attached on a tuning fork quartz crystal resonator (TF-QCR). A PZT transducer is employed to vibrate the assembly, which is composed of the TF-QCR, the micro-stylus, and the micro-sphere at the resonance frequency of the TF-QCR. The variation of the vibration frequency is detected by the TF-QCR based on the piezoelectric effect of the quartz crystal. The vibration of the TF-QCR is detected and amplified by current-to-voltage (I-V) converter. The output of the I-V converter is then utilized to keep the vibration frequency constant by using a self-excitation circuit that consists of a phase locked loop circuit (PLL). It should be noted that the actual resonance frequency of the TF-QCR used in the assembly is confirmed to be approximately 30 kHz, which is smaller than that of the single TF-QCR element of 32.768 kHz, due to the influence of the masses of the micro-stylus and the micro-sphere attached on the TF-QCR. The vibration direction is set in parallel with the measuring surface.

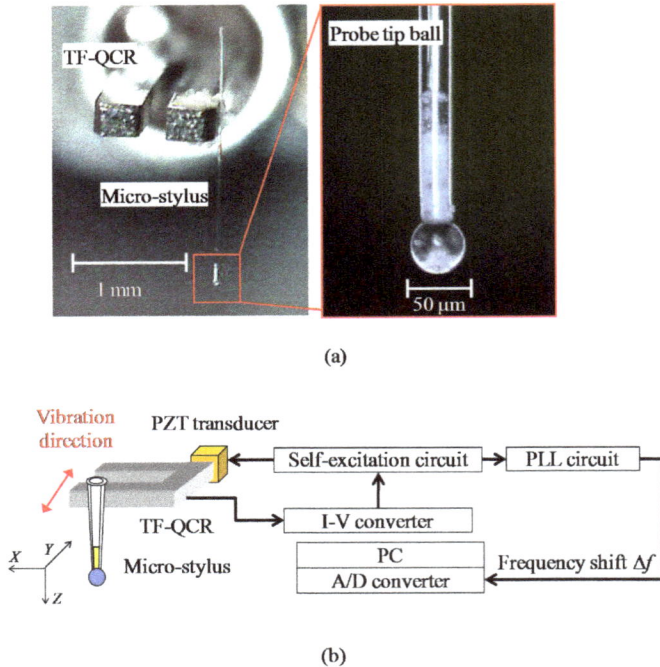

(a)

(b)

Figure 1. Configuration of shear-mode micro-probe. (**a**) Photograph of shear-mode micro-probe; (**b**) Schematic diagram of shear-mode micro-probe.

To determine the trigger point for the probing of the micro-probe to the measured surface, the interaction force of the water layer on the measured surface against the probe is detected through detecting the change of the probe vibration with the TC-QCR [26]. When the tip ball of the vibrating probe contacts with the water layer of the measured surface, the frequency of the probe vibration is shifted due to the adhesion force of the water layer. The output of the TF-QCR is input to a phase locked loop (PLL) circuit for detection of the amount of frequency shift. The PLL circuit output is then employed as the trigger signal for the probing of the micro-probe to the measured surface. The PLL output signal is fed into analog-to-digital (A/D) converter of a personal computer (PC). A certain frequency shift amount is set to be a threshold of the trigger signal.

Figure 2 shows a schematic diagram of the micro-CMM by using the shear-mode micro-probe. The probe is mounted on the PZT-driven linear stages (P-621.2CL and P-621.1ZCL, PI), which can be moved in the X- and Z-directions independently, for fine positioning of the probe. The stroke and resolution of the PZT stages are 100 μm and 1 nm, respectively. The positioning accuracy in the X- and Z-directions are ±2 nm and ±1 nm, respectively. Since the gap width of the slot die is calculated based on the X-directional probe displacement, a commercial laser interferometer (RLE10, RENISHAW) with a resolution of 0.39 nm is employed for the measurement of the X-directional probe displacement. A plane mirror is mounted on the opposite side of the probe on the X-direction PZT stage as a moving mirror for the laser interferometer. In order to reduce the influence of the Abbe error, the optical axis of the laser interferometer is aligned with the center of the probe tip. The PZT stages are mounted on the DC servo motor driven positioning stages (M-111.1DG, PI) for coarse positioning of the probe in the X- and Z-directions. The stroke and resolution of the servo motor coarse stages are 25 mm and 6.9 nm, respectively. The positioning accuracy resolution of the servo motor stages are 0.1 μm. For positioning the probe in the Y-direction, an air bearing linear slide (TAAT30SL-19+C, NTN) with a travel stroke of 300 mm is employed. The resolution and positioning accuracy of the Y-directional linear slide are 20 nm and ±0.1 μm, respectively. The precision slot die with a length of 200 mm is placed on the moving table of the Y-directional linear slide for measurement.

When a typical CMM is employed for measuring the gap distance between two parallel surfaces, at least three points on one surface are probed first to define the surface. A point on the other surface is then probed for obtaining the coordinates of the point. The gap distance between the two surfaces can thus be accurately obtained from the coordinates of the point on the second surface with respect to the defined first surface even if the normal of the surfaces are not parallel/vertical to the coordinate axes of the CMM. However, it is necessary to conduct the probing at least three points on the first surface, which is a time-consuming process. In order to shorten the measurement time, the gap width of the precision slot die is measured in the micro-CMM, as shown in Figure 3, by probing only one point on each of the surfaces forming the micro-slit under the condition that the axes of the X-directional positioning stages are carefully aligned to be parallel to the normal of the slit surfaces and perpendicular to the Y-axis of the linear slide.

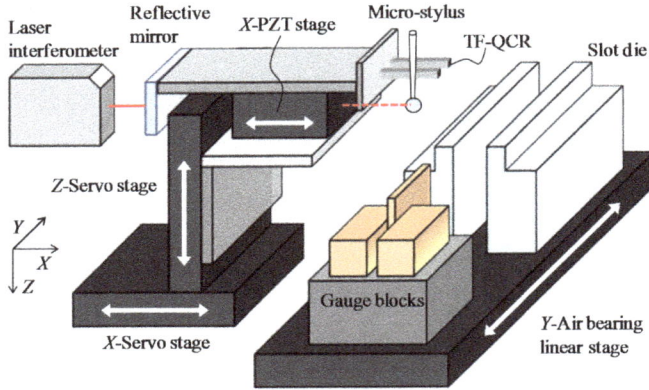

Figure 2. Experimental configuration of the micro-CMM (micro-coordinate measurement machine) by using a shear-mode micro-probe.

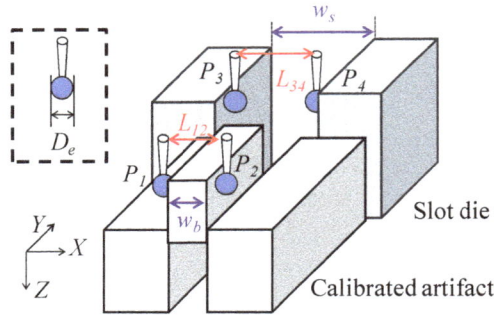

Figure 3. Schematic diagram of on-line qualification and length measurement of gap width.

Consequently, the gap width of the slot die w_s can be calculated by the following equation.

$$w_s = L_{34} + D_e + 2s_{ts} \tag{1}$$

where L_{34} is the distance between the trigger positions P_3 and P_4 along the X-direction. D_e is the effective diameter of the probe tip ball. s_{ts} is the thickness of the water layer on the slot die at the trigger point. The effective diameter D_e is influenced by the water layer on the measured surface, the stiffness of the stylus shaft, and so on. The probing can be carried out by the shear-mode micro-probe through detection of the interaction between the water layer on the measured surface and that on the

118

surface of the probe tip ball. Similar to the authors' previous research [32], on-line qualification of the effective diameter D_e is carried out soon before the gap width measurement to reduce the combined standard uncertainty caused by the variations in the water layer thickness of the water layer, which is dependent on measurement conditions such as temperature, humidity, and surface materials. Firstly, the effective diameter of the probe tip is estimated by using a calibrated artifact. A grade K gauge block (Mitsutoyo) with a thickness of 100 μm and a tolerance of ±10 nm is used as the calibrated artifact. The tolerance of the gauge block was determined based on the calibration certification. Both the slot die and the gauge block are made of stainless steel. The effective diameter D_e can be estimated by:

$$D_e = L_{12} - w_b - 2s_{tb} \tag{2}$$

where w_b is the thickness of the calibrated artifact, L_{12} is the distance between the trigger points P_1 and P_2 along the X-direction, and s_{tb} is the thickness of the water layer on the calibrated artifact.

The stroke of the PZT stage used in the micro-CMM is 100 μm. It is not enough to move the probe across the calibrated artifact for probing the two sides of the artifact. The X-servo motor stage is therefore used to move the probe to the opposite side of the artifact for the probing at P_2 after the probing at P_1. In order to prevent the motion error induced by the Z-servo motor stage, it is desired to move the probe in the X-direction without lifting the probe up in the Z-direction. For this purpose, the artifact is moved by the Y-slide table in the negative Y-direction until the moving path of the probe along the X-direction is not blocked by the artifact. After the probe is moved across the artifact by the X-servo motor stage, the artifact is moved back to its original position by the Y-slide table in the positive Y-direction. Finally, the probing at P_2 is carried out by using the X-directional PZT stage. As a result, Equation (1) can be rewritten as follows.

$$w_s = L_{34} + D_e + 2s_{ts} = L_{34} + L_{12} - w_b - 2\left(s_{tb} - s_{ts}\right) \tag{3}$$

Since s_{tb} is almost the same as s_{ts}, the influence of the water layer thickness can be cancelled. Figure 4 shows the vibration frequency shifts when probing the slot die and the gauge block, respectively. The vertical axis in Figure 4 indicates the frequency shift Δf and the horizontal axis indicates the probe displacement in the X-direction. When the frequency shift is extended to the trigger threshold, the probe position in the X-direction is measured by the laser interferometer to determine the trigger point of the probing. The probe tip is then retracted back from the measured surface immediately. As shown in Figure 4, if the frequency shift Δf of 0.1 Hz is set to be the trigger threshold for the probing, the difference of the probe displacements for the two results in Figure 4 is less than 2 nm. Consequently, the gap width of the slot die can be measured accurately by the on-line qualification of the effective diameter of the tip ball.

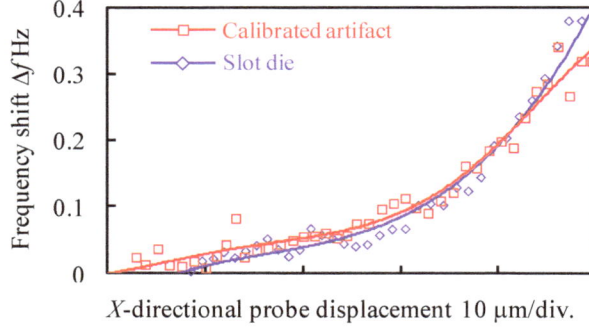

X-directional probe displacement 10 μm/div.

Figure 4. Vibration frequency shifts when probing the calibrated artifact and the slot die.

After the on-line qualification procedure, the probe tip is positioned inside of the micro-slit of the slot die for the gap width measurement based on Equation (1). The gap width uniformity can also be obtained by measuring the gap widths at different Y-directional positions through moving the slot die with the Y-directional linear slide.

3. Experimental Result and Discussion

3.1. Evaluation of Alignment Errors of the Measurement System

As described above, the gap width measurement is carried out in the developed micro-CMM by probing one point on each of the surfaces of the micro-slit. The reliability of the measurement results is influenced by the angular alignment errors and motion errors of the moving stages. Figure 5 shows a schematic of the angular alignment errors of the moving stages. The moving axis of the air bearing linear slide is defined as the Y-axis. The X-axis and Z-axis are shown in Figure 5. The length direction of the micro-slit of the slot die and the surface of the gauge block are arranged in parallel with the Y-axis. The angular alignment errors of the slot die and the calibrations around the Z-axis are indicated as θ_{gb} and θ_{sd}, respectively. θ_{gb} and θ_{sd} are calculated based on the X-directional deviation Δd_x obtained by moving the Y-directional linear slide. Δd_x is measured by using an optical fiber displacement sensor with a resolution of 0.49 nm. Therefore, θ_{gb} and θ_{sd} can be calculated by using the following equations.

$$\theta_{gb} = \tan\left(\frac{\Delta d_{x\_gb}}{L_{gb}}\right) \tag{4}$$

120

$$\theta_{sd} = \tan\left(\frac{\Delta d_{x\_sd}}{L_{sd}}\right) \tag{5}$$

where Δd_{x_gb} and Δd_{x_sd} are the X-directional displacements of the gauge block and the slot die due to the tilt around the Z-axis of the slot die, respectively. L_{gb} and L_{sd} are the Y-directional displacements caused by the linear slide, respectively. L_{gb} and L_{sd} are set to be 180 mm and 20 mm, respectively.

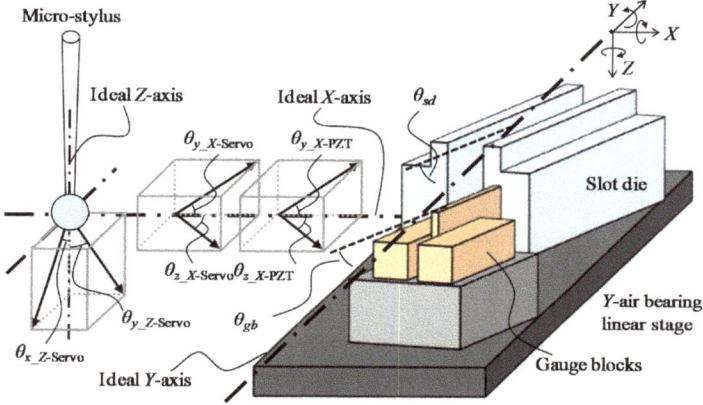

Figure 5. Schematic of the angular alignment errors of the moving axes of the measurement system.

The alignment errors of the X-directional moving axis of the PZT stage around the Y-axis and the Z-axis are indicated as θ_{y_X-PZT} and θ_{y_X-PZT}, respectively. Figure 6a shows a schematic diagram of the evaluated angular alignment errors of the X-directional stages. A right-angle prism mirror is placed on the table of the Y-linear slide for measurement of the deviations in the Y- and Z-directions with the optical fiber displacement sensor. The Y-Z plane of the prism mirror is aligned in parallel with the moving axis of the Y-linear slide. Based on Equations (4) and (5), θ_{y_X-PZT} and θ_{y_X-PZT} can be calculated by moving the X-directional PZT stage. Similarly, for the X-direction servo motor stages, the alignment errors around the Y-axis and Z-axis are defined as $\theta_{\_X-Servo}$ and $\theta_{z\_X-Servo}$, respectively.

The Z-directional servo motor stage is used to adjust the measurement position inside the micro-slit along the Z-direction. The alignment errors of the Z-directional servo motor stage around the X-axis and the Y-axis are indicated as $\theta_{x\_Z-Servo}$ and $\theta_{y\_Z-Servo}$, respectively. As shown in Figure 6b, $\theta_{x\_Z-Servo}$ and $\theta_{y\_Z-Servo}$ can be evaluated by using the right-angle prism mirror and the optical fiber displacement sensor.

121

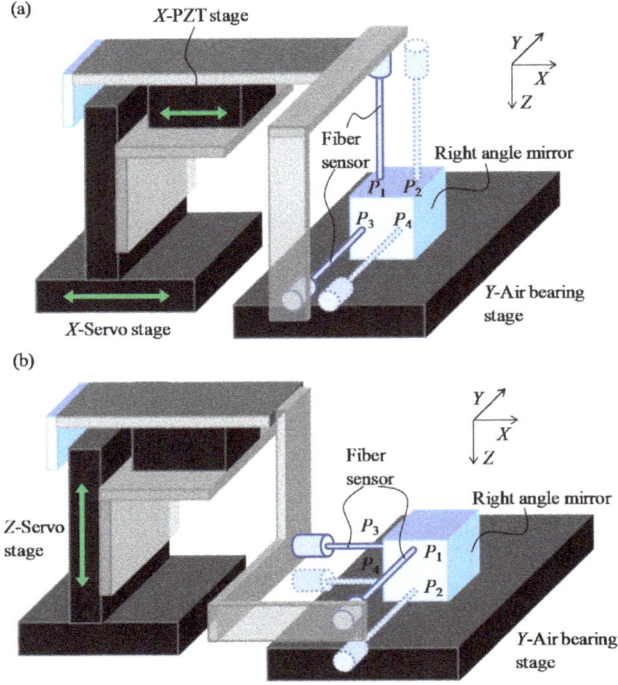

Figure 6. Schematic for evaluation of the angular alignment errors: (**a**) X-directional stages; (**b**) Z-directional stage.

Figure 7 shows a schematic of the tilt errors of the reflective mirror with respect to the optical axis of the laser interferometer. Since the gap width of the micro-slit is obtained from the outputs of the laser interferometer, the alignment error of the laser interferometer is also an uncertainty source for the gap width measurement. When the reflective mirror of the interferometer, which is mounted on the opposite side of the micro-probe, is not perpendicular to the X-axis, a cosine error will occur, as shown in Figure 7. θ_{y_mirror} and θ_{z_mirror} are defined as the tilt errors around the Y- and Z-axes with respect to the optical axis, respectively. Figure 8 shows a schematic of the cosine error between the laser interferometer and X-directional moving axis. L is the distance measured by the laser interferometer. L' is the actual displacement of the probe tip along the X-direction. L'/L can be expressed by the following equation.

$$\frac{L'}{L} = \frac{1}{\cos \theta_{y\_mirror} \cos \theta_{z\_mirror}} \tag{6}$$

In order to evaluate the alignment errors of the reflective mirror with respect to the laser interferometer, the tilt errors of the reflective mirror were measured by using the optical fiber displacement probe. Figure 9 shows a schematic diagram for evaluation of the alignment errors of the reflective mirror. One plane of the right-angle prism mirror is arranged perpendicular to the X-axis of the PZT stage. The difference of the tilting angles between the prism mirror and the reflective mirror is measured. Table 1 shows a summary of the evaluated angular alignment errors of the developed micro-CMM.

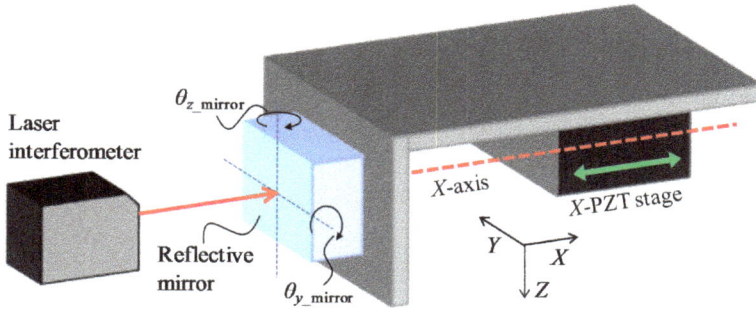

Figure 7. Tilt errors of the reflective mirror with respect to the optical axis of the laser interferometer.

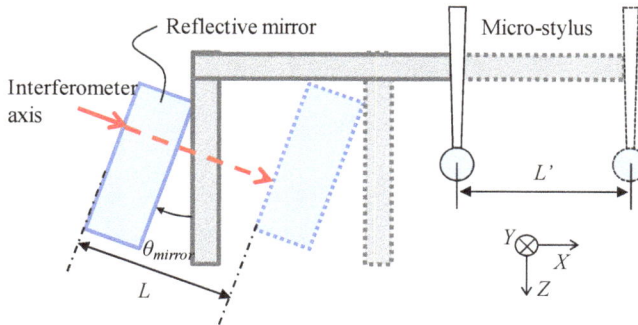

Figure 8. Schematic of the cosine error between the laser interferometer and the X-directional moving axis.

Table 1. Summary of the angular alignment errors of the positioning stages.

$\theta_{y\_X\text{-}PZT}$	$\theta_{z\_X\text{-}PZT}$	$\theta_{y\_X\text{-}Servo}$	$\theta_{z\_X\text{-}Servo}$	$\theta_{x\_Z\text{-}Servo}$
7.5 mrad	2.8 mrad	14.0 mrad	3.6 mrad	5.7 mrad
$\theta_{y\_Z\text{-}Servo}$	$\theta_{Slotdie}$	$\theta_{Gauge\text{-}block}$	θ_{y_mirror}	θ_{z_mirror}
14.1 mrad	0.03 mrad	0.07 mrad	6.8 mrad	0.34 mrad

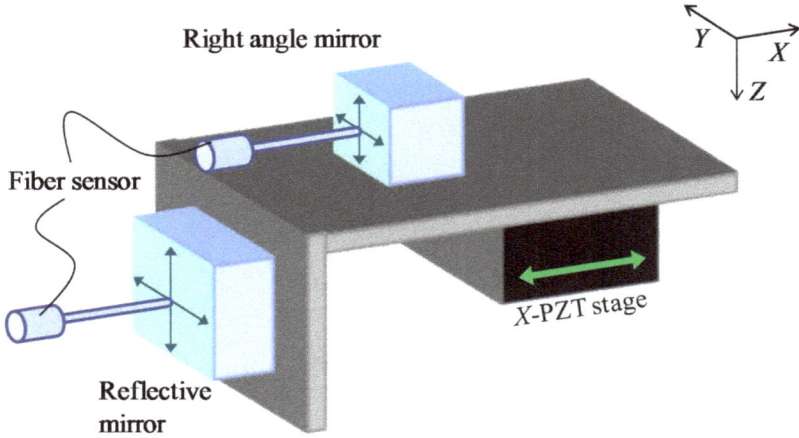

Figure 9. Schematic of angular alignment errors evaluation of the reflective mirror of the laser interferometer.

Based on the evaluated alignment errors of the positioning stages, the geometric measurement errors are investigated. Figure 10 shows a geometrical model in the X-Y plane during the probing on the calibrated artifact. L_{gb} is the distance between the probing points on the calibrated artifact. h is the distance between the moving axes of the X-directional PZT stage on the both sides of the gauge block. D_s is the X-directional displacement of the servo motor stage. The thickness of the calibrated artifact w_b can be expressed by the following equation.

$$w_b = \left(L_{gb} - L'\right) \cos\left(\theta_{z\_X-PZT} - \theta_{gb}\right) \tag{7}$$

L' can be defined as follows.

$$L' = h \tan\left(\theta_{z\_X-PZT} - \theta_{gb}\right) \tag{8}$$

h can be calculated by

$$h = D_s \sin\left(\theta_{z\_X-Servo} - \theta_{z\_X-PZT}\right) \tag{9}$$

124

Therefore, the measurement error Δw_b of the calibrated artifact width, which is also extended in the Y-Z plane, can be calculated by the following equation.

$$
\begin{aligned}
\Delta w_b &= \left(L_{gb} - D_s \right) - w_b \\
&= \left(\frac{w_b + D_s \, \tan\left(\theta_{y\_X-servo} - \theta_{y\_X-PZT}\right)\tan\left(\theta_{y\_X-PZT} - \theta_{gb}\right)}{\cos\left(\theta_{z\_X-PZT} - \theta_{gb}\right)} - w_b \right) \\
&+ \left(\frac{w_b + D_s \, \tan\left(\theta_{z\_X-servo} - \theta_{z\_X-PZT}\right)\tan\left(\theta_{z\_X-PZT} - \theta_{gb}\right)}{\cos\left(\theta_{z\_X-PZT} - \theta_{gb}\right)} - w_b \right)
\end{aligned}
\tag{10}
$$

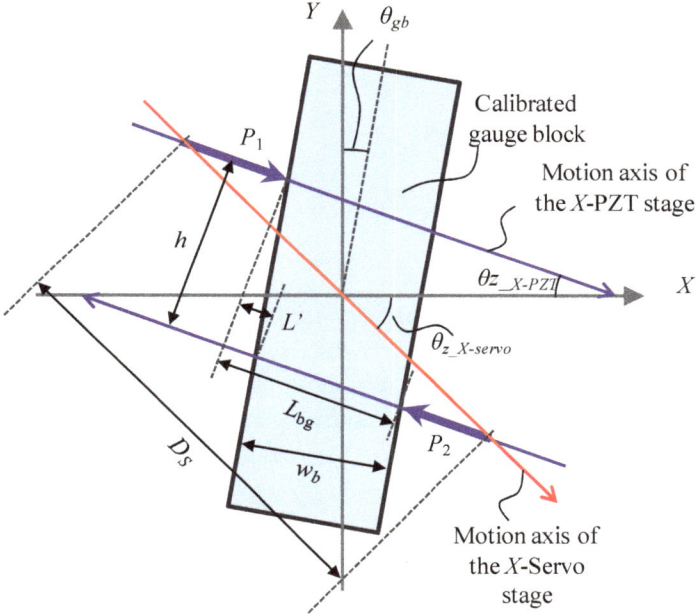

Figure 10. Geometrical model of the alignment error of the probing points during the qualification procedure.

Figure 11 shows a geometrical model in the X-Y plane. The measurement error Δw_s of the slit gap width can be calculated by the following equation.

$$
\Delta w_s = \frac{w_s}{\cos\left(\theta_{y\_X-PZT} - \theta_{y\_sd}\right)\cos\left(\theta_{z\_X-PZT} - \cos\theta_{z\_sd}\right)} - w_s
\tag{11}
$$

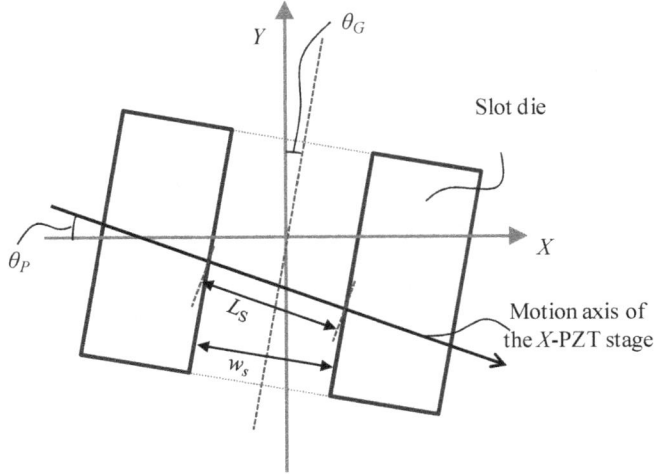

Figure 11. Geometrical model of the alignment error of the probing points during the gap width measurement.

Since the gap width measurement is carried out by the probe displacement with the laser interferometer in the X-direction, the Abbe errors for the laser interferometer have to be considered for the evaluation of the combined standard uncertainty. When the center of the probe tip is not located on the optical axis of the laser interferometer, an Abbe error component will be caused in the measurement result of the probe displacement due to the tilt errors of the X-directional stages. Figure 12 shows the Abbe error between the probe tip and the laser interferometer. As shown in Figure 12, a_1 indicates the X-directional distance between the probe tip and the reflective mirror. a_2 is the Z-directional difference between the center of the probe tip and that of the laser spot on the reflective mirror. θ_{Abbe} is the tilt error of the X-directional stages. The Abbe errors L_{y_Abbe} and L_{z_Abbe} can be expressed as follows.

$$L_{y\_Abbe} = \left(a_1 \tan \frac{\theta_{y\_X-PZT}}{2} + a_2\right) \tan \theta_{y\_X-PZT} + \left(a_1 \tan \frac{\theta_{y\_X-Servo}}{2} + a_2\right) \tan \theta_{y\_X-Servo} \quad (12)$$

$$L_{z\_Abbe} = \left(a_1 \tan \frac{\theta_{z\_X-PZT}}{2} + a_2\right) \tan \theta_{z\_X-PZT} + \left(a_1 \tan \frac{\theta_{z\_X-Servo}}{2} + a_2\right) \tan \theta_{z\_X-Servo} \quad (13)$$

Consequently, the combined Abbe error can be expressed in the following equation.

$$L_{Abbe\_error} = \sqrt{L_{y\_Abbe}^2 + L_{z\_Abbe}^2} \quad (14)$$

The measurement error for the gap width measurement caused by the tilt errors of the positioning stages is then evaluated. When the probing points on the slot die surface are changed by moving the slot die with the Y-linear slide, errors will be introduced in the measurement result of the gap width by the tilt error about the

126

Y-axis (Rolling) $\theta_{slide\_rolling}$ and the tilt error about the X-axis (Yawing) θ_{slide_yawing}. The gap width measurement error Δw_{y_slide} caused by the tilt errors of the Y-linear slide can be indicated as follows.

$$\Delta w_{y\_slide} = \frac{w_s}{\cos\left(\theta_{slide\_rolling}\right)\cos\left(\theta_{slide\_yawing}\right)} - w_s \tag{15}$$

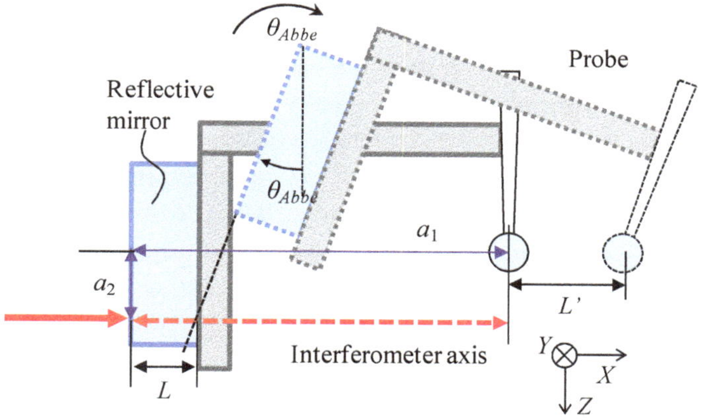

Figure 12. Schematic of Abbe error caused by the tilting motion of X-directional moving stage.

When the probing points are changed by moving the slot die with the Z-directional servo motor stage, errors will be introduced to the measurement result of the gap width by the tilt error $\theta_{z\text{-}servo\_rolling}$ about the Y-axis (Rolling) and the tilt error $\theta_{z\text{-}servo\_pitching}$ about the X-axis (Yawing). The gap width measurement error $\Delta w_{z\text{-}servo}$ caused by the tilt errors of the Y-linear slide can be indicated as follows.

$$\Delta w_{z-servo} = \frac{w_s}{\cos\left(\theta_{z-servo\_rolling}\right)\cos\left(\theta_{z-servo\_pitching}\right)} - w_s \tag{16}$$

Table 2 shows a summary of the tilt errors of the positioning stages. The tilt errors are measured by a laser autocollimator with a resolution of 1.21 μrad.

The measurement uncertainties for on-line qualification and gap width measurement can then be evaluated based on Equations (4)–(15) as well as Tables 1 and 2.

Table 2. Summary of tilting errors of the positioning stages.

Stage	Axis	Stroke	Tilting Error
Y-linear slide	Rolling	300 mm	8.10 mrad
	Yawing		12.31 mrad
Z-servo motor	Rolling	1 mm	22.59 mrad
	Pitching		73.40 mrad

3.2. Experiments of On-Line Qualification and Gap Width Measurement

The micro-slit on the precision slot die for measurement has a nominal gap width of 85 ± 5 µm over a length of 200 mm. To avoid the influence of the chamfered edges of the precision flat parts forming the micro-slit, it is necessary for the probe tip to be arranged inside of the micro-slit for measurement of the gap width. In the developed micro-CMM, the probing position along the Z-direction is determined by utilizing the measurement capability of the shear-mode micro-probe in both the X- and Z-directions [26]. Figure 13 shows the detection strategy and the detection results of the chamfered edges of the two precision flat plates, respectively. It can be seen that the inner surfaces of the micro-slit can be identified from the measurement result shown in Figure 13b. The Z-directional height of the chamfered edges of the precision flat plates was estimated to be approximately 90 µm and the probing position along the Z-direction was selected to be lower than areas of the chamfered edges for the following experiments of on-line qualification and gap width measurement.

On-line qualification of the effective diameter of the probe tip ball was then conducted by using the calibrated artifact. As a set of the on-line qualification operation, the probe was moved to the side of P_2 by the X-directional servo motor stage for another 5 times of probing at P_2 after 5 times of probing at P_1. It took approximately 50 s for the set of the operation. The same set of the operation was repeated 5 times to evaluate the repeatability of probing at the trigger points. Figure 14 shows the results of the on-line qualification. The X-directional positions of the trigger points at P_1 and P_2, which is a contact points between the probe tip ball and the inside surface of the slot die, are shown in Figure 14a,b, respectively. The vertical axes shown in Figure 14a,b show the X-directional probing positions calculated by the output of the laser interferometer. The mean values of the X-directional positions of the trigger points at P_1 and P_2 over the 5 sets of on-line qualification operations were 5.625 µm and 159.344 µm, respectively. Due to the thermal drift between the laser interferometer and the reflective mirror mounted on the X-PZT stage, the mean values at P_1 and P_2 were changed almost linearly due to the influence of thermal drifts. Figure 14c shows the value of L_{12} calculated by taking the difference of the corresponding results of Figure 14a,b, in which the thermal drift error components were removed. The mean and the standard deviation of L_{12} over the 5 sets of on-line qualification were 153.781 μm and 11.5 nm, respectively. Consequently, the effective diameter of the probe tip ball D_e was estimated to be 53.781 μm.

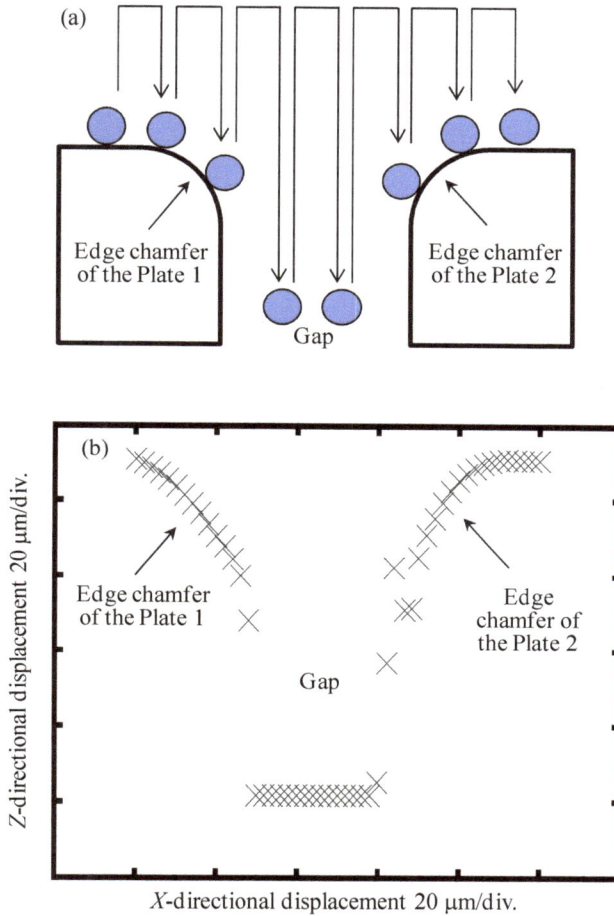

Figure 13. Chamfered edge detection by using the micro-probe: (**a**) Detection strategy of the chamfered edge; (**b**) Measurement result of the slit position.

After the on-line qualification, the probing position was located to the inside of the micro-slit by moving the slot die with the Y-linear slide for the gap width measurement. At each Y-directional measurement position, after the probing at P_3, the probe was moved by the X-directional PZT stage toward the opposite surface of the micro-slit for the probing at P_4. This operation was repeated 5 times. The results are shown in Figure 15. Figure 15a,b show the probing positions at P_3 and P_4, which indicate the X-directional probe contact positions at the inner surface of the slot die. The mean values at P_3 and P_4 are -13.392 μm and 13.865 μm, respectively.

Figure 15c shows the values of L_{34} calculated based on the results in Figure 15a,b. The mean and the standard deviation of L_{34} were estimated to be 27.243 μm and 14.8 nm, respectively. According to Equation (3), the gap width of the slot die micro-slit was estimated to be 80.962 μm.

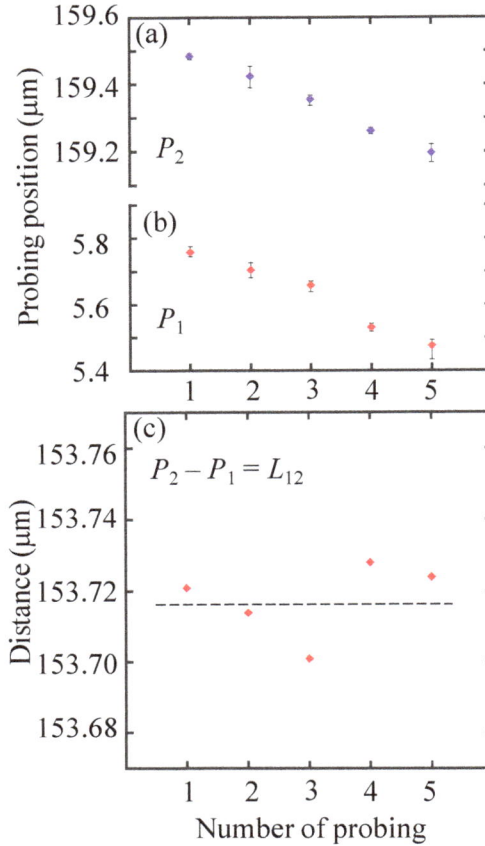

Figure 14. Experimental results of on-line qualification procedure: (**a**) Probing results at P_2; (**b**) Probing results at P_1; (**c**) Measurement results of L_{12}.

In order to evaluate the accuracy of the gap width measurement result, uncertainty analysis was carried out for each of the terms in Equation (3) based on GUM (ISO *Guide to the Expression of Uncertainty in Measurement*) [33]. Table 3 shows a summary of the uncertainty budget for the gap width measurement. u_{ws}, which was the combined standard uncertainty of L_{34}, is calculated based on Figure 11, Equations (11) and (14). The repeatability of probing was one of the main uncertainty sources for u_{ws}. The total time for the gap width measurement was approximately

9) s, during which the temperature was measured to be 22.677 °C $\pm$ 0.011 °C. Consequently, u_{ws} is estimated to be 10.5 nm, which corresponds to the combined standard uncertainty for the measurement of L_{34}.

Table 3. Uncertainty budget (unit: nm).

Uncertainty sources	Symbol	Value	Coverage Factor	Standard Uncertainty
Uncertainty in w_S	u_{ws}	-	-	10.5
Cosine error by the alignment of the gauge block and the probing axis	$u_{cos\_slotdie}$	2.6	$\sqrt{3}$	1.7
Cosine error by the alignment of the interferometer axis and the probing axis	u_{cos_laser}	0.6	$\sqrt{3}$	0.3
Abbe error of the X-PZT stage	u_{pzt_abbe}	5.27	$\sqrt{3}$	3.0
Resolution of the interferometer	$u_{laser\_resolution}$	0.79	$\sqrt{3}$	0.5
Linearity error of the interferometer	$u_{laser\_linearity}$	5.0	$\sqrt{3}$	2.9
Thermal drift by temperature change	$u_{Thermal\_drift}$	7.0	$\sqrt{3}$	4.0
Repeatability of L_{34}	u_{rep_L34}	14.8	$\sqrt{3}$	8.5
Uncertainty in D_e	u_{De}	-	-	26.4
Cosine error by the alignment of the gauge block and the probing axis	u_{cos_gauge}	11.0	$\sqrt{3}$	6.4
Cosine error by the alignment of the interferometer axis and the probing axis	u_{cos_laser}	3.58	$\sqrt{3}$	2.1
Abbe error of the X-PZT stage	u_{pzt_abbe}	5.27	$\sqrt{3}$	3.0
Abbe error of the X-servo stage	u_{servo_abbe}	41.5	$\sqrt{3}$	24.0
Resolution of the interferometer	$u_{laser\_resolution}$	0.79	$\sqrt{3}$	0.5
Linearity error of the interferometer	$u_{laser\_linearity}$	5.0	$\sqrt{3}$	2.9
Thermal drift by temperature change	$u_{Thermal\_drift}$	7.0	$\sqrt{3}$	4.0
Repeatability of L_{12}	u_{rep_L12}	11.5	$\sqrt{3}$	6.6
Uncertainty in w_b	u_{wb}	-	-	17.3
Length tolerance(Calibrated gauge block)	$u_{tol\_calibrate}$	20.0	$\sqrt{3}$	17.3
Uncertainty in Ds	u_s	-	-	1.2
Uncertainty due to water layer	u_{water}	2.0	$\sqrt{3}$	1.2
Expanded uncertainty (with a coverage factor $k = 2$)	U	-	-	66.6

u_{De}, which was the combined standard uncertainty of the on-line qualification of the effective diameter of the probe tip ball, was estimated based on Figure 10, Equations (10) and (14). As shown in Table 3, the cosine error caused by the misalignment of the gauge block and the probing axis was a relatively large uncertainty source because the X-directional servo stage was moved during the on-line qualification process. The total time for the on-line qualification of the effective diameter was approximately 250 s, during which the temperature was measured to be 22.677 °C $\pm$ 0.008 °C. As a result, u_{De} was estimated to be 26.4 nm. u_{wb} was the length tolerance of the calibrated gauge block. According to the calibration certification of the gauge block used in the micro-CMM, u_{wb} was estimated to be 17.3 nm. $u_{\Delta s}$ was the uncertainty source introduced by the water layer on the measured surface. It was estimated to be 1.2 nm, based on Figure 4. Consequently, the expanded uncertainty U of the gap width measurement was estimated to be

66.6 nm ($k = 2$), which was smaller than the allowed maximum expanded uncertainty of 100 nm.

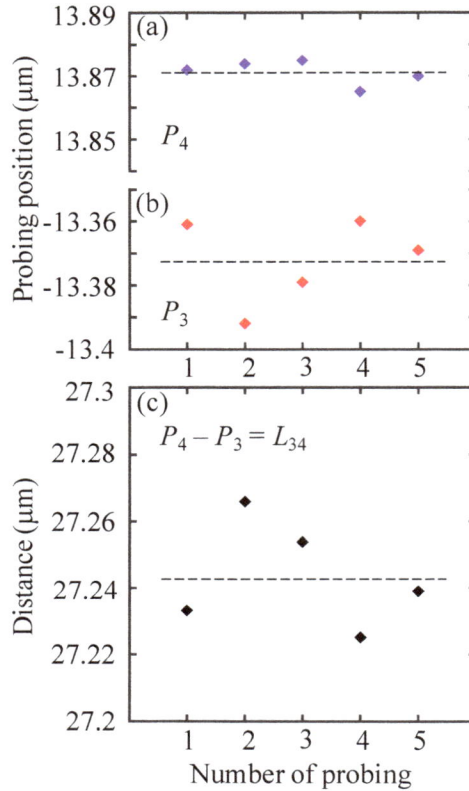

Figure 15. Experimental results of gap width measurement: (**a**) Probing results at P_4; (**b**) Probing results at P_3; (**c**) Measurement results of L_{34}.

By changing the probing position along the Y- and Z-directions, the gap width uniformity was measured. Figure 16 shows the gap widths at measurement positions with an equal interval of 10 mm along the Y-direction. The measurement depth was set to be 50 μm. The probing was repeated 3 times at each measurement position. The dot plot in the figure shows the mean value of the repeated probing results and the error bar represents the corresponding standard deviation at each measurement position. The maximum and minimum values of the measured gap widths along the Y-direction were 78.883 μm and 81.980 μm, respectively. The total measurement time was approximately 1800 s. Since the two precision flat plates of the slot die were fastened by the screws at the center and the two ends of the slot die, the gap width tended to be smaller at the positions of the screws, which was corresponding to the measurement results in Figure 16. The maximum and minimum standard deviations of the measured gap widths were 27 nm and 4 nm, respectively. The employment of the long stroke air-bearing linear slide made it possible for measurement of the gap width uniformity over the entire slot die length of 200 mm, which was a significant improvement on a conventional micro-CMM.

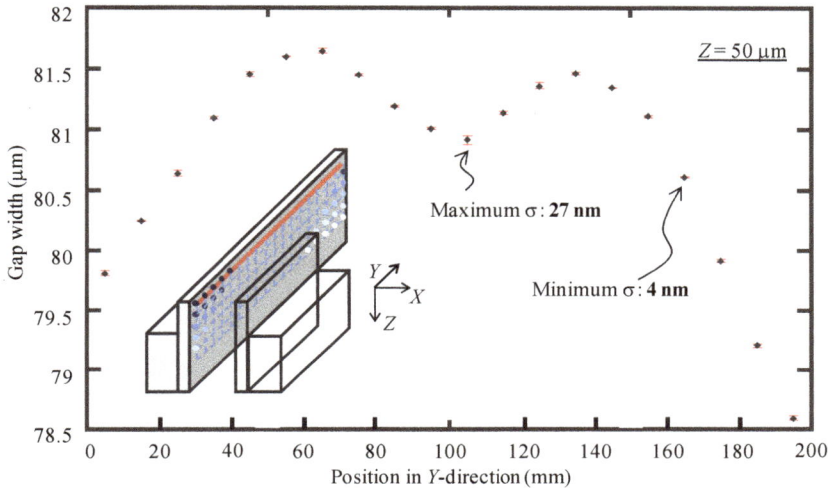

Figure 16. Measurement results of gap uniformity of slot die along the Y-direction.

Figure 17 shows the measured gap widths at the measurement positions with an equal interval of 100 μm along the Z-direction. The measurements were carried out at $Y = 100$ mm. The probing was repeated 5 times at each measurement position. The dot plot in the figure shows the mean value of the repeated probing results and the error bar represents the corresponding standard deviation at each measurement position. The maximum and minimum standard deviations of the measured gap widths were 27 nm and 4 nm, respectively. The total measurement time was approximately 990 s. The gap widths along the Z-direction were almost uniform. It can be seen from the figure that the gap width was successfully measured at $Z = 1000$ μm, indicating the effective working distance of the developed micro-CMM along the Z-direction.

The gap widths along different Y-directional lines were measured by changing the measurement depths along the Z-direction. The interval between the measurement points along the Y-direction was 10 mm and that along the Z-direction was 100 μm. The total number of probing points was 220. During the measurement, the temperature was measured to be 24.1 ± 0.1 °C. The total measurement time was approximately 200 min. As can be seen in Figure 18, the gap widths along different Y-lines were quite similar to each other. The measured gap widths at the same Y-position but different Z-positions were employed to demonstrate the gap

133

widths at different Z-lines. The results are shown in Figure 19. It can be seen that the gap width slightly decreased with the increase in the measurement depth for all the Z-lines.

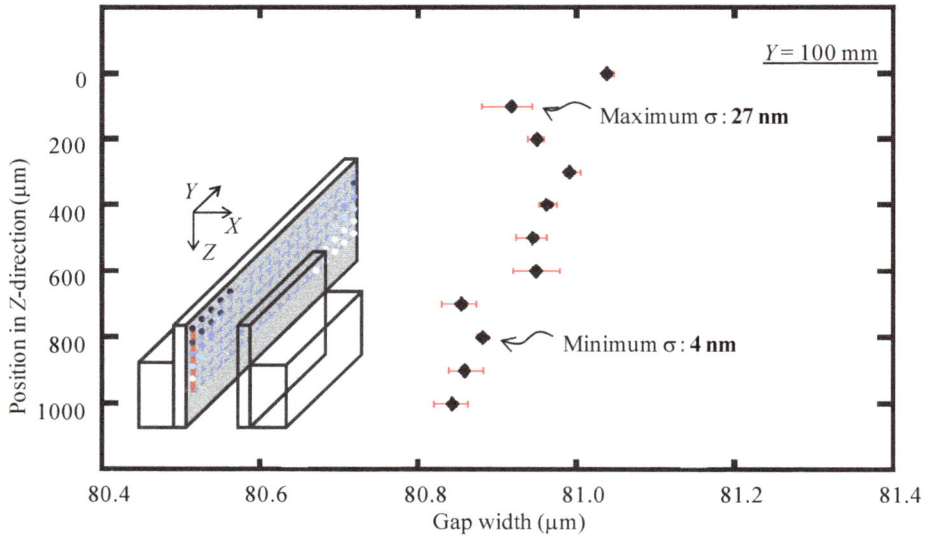

Figure 17. Measurement results of gap uniformity of slot die along the Z-direction.

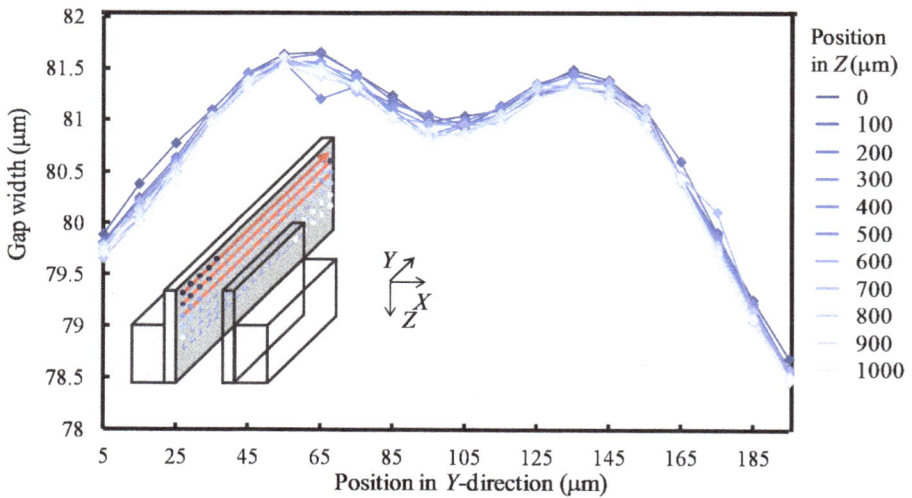

Figure 18. Gap width uniformity over the entire slot die length (Y-direction).

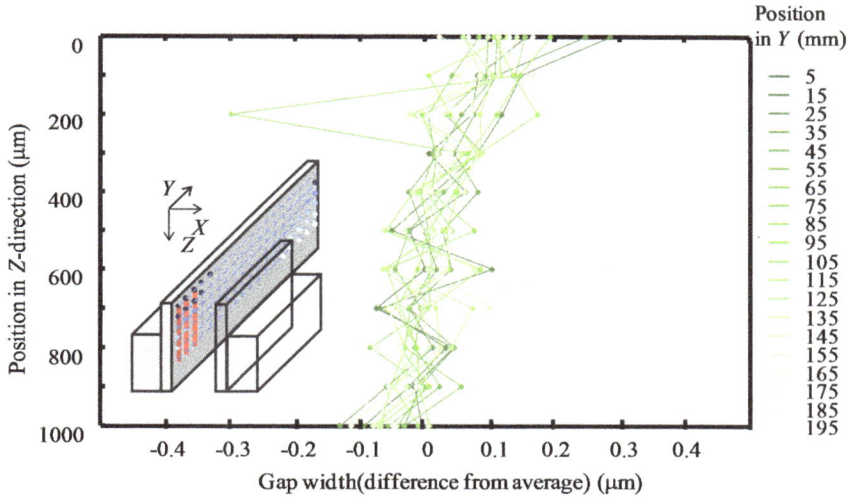

Figure 19. Gap width uniformity over the entire slot die length (Z-direction).

4. Conclusions

Dimensional measurement of micro-slit gap width of a precision slot die coater with a length of 200 mm and a nominal width of 85 μm has been carried out by using a specially designed and constructed micro-CMM. A shear-mode micro-probing system was employed in the micro-CMM. A calibrated artifact, which was a calibrated gauge block with a thickness of 100 μm, was placed on the same stage table as a Y-directional linear slide for the on-line qualification of the effective diameter of the probe tip ball prior to the gap width measurement of the slot die. The proposed micro-CMM, which combined the micro-probing system and a laser interferometer, was able to achieve nanometric resolution and repeatability of length measurement. The effective diameter D_e of the probe tip ball was evaluated to be 53.781 μm, and the expanded uncertainty in D_e was estimated to be 52.8 nm ($k = 2$). After the qualification procedure, the gap width measurement of the slot die was carried out immediately. As a result, the expanded uncertainty of the gap width measurement was estimated to be 66.6 nm ($k = 2$), which can satisfy the required measurement accuracy. The use of an air-bearing linear slide with a stroke of 300 mm has made it possible for the micro-CMM to cover the entire length of the precision slot die. Furthermore, a micro-stylus with an effective working length of larger than 1 mm, which was composed of a capillary glass shaft and a micro-glass sphere, has been employed to measure the gap width from the inside of the micro-slit of the precision slot die.

On the other hand, the shear-mode micro-probe was utilized as a touch-trigger probe in the developed micro-CMM and the total measurement time for the uniformity evaluation of the gap width was approximately 200 min. Reduction of the measurement time will be an important future work.

Acknowledgments: This research was supported by KAKENHI. The authors would like to acknowledge the Ministry of Education, Culture, Sports, Science and Technology (MEXT) and Japan Society for the Promotion of Science (JSPS).

Author Contributions: S.I. and W.G. conceived the experiments and wrote the paper; H.K. designed the experimental apparatus and performed the experiment; Y.C and Y.S. analyzed the data. K.T., T.K., K.A. and A.H. contributed to prepare slot die sample.

Conflicts of Interest: The authors declare no conflict of interest.

References

1. Chen, Y.L.; Gao, W.; Ju, B.F.; Shimizu, Y.; Ito, S. A measurement method of cutting tool position for relay fabrication of microstructured surface. *Meas. Sci. Technol.* **2014**, *25*, 064018.
2. Aziz, M.; Ohnishi, O.; Onikura, H. Novel micro deep drilling using micro long flat drill with ultrasonic vibration. *Precis. Eng.* **2012**, *36*, 168–174.
3. Hao, T.; Yong, L.; Long, Z.; Baoquan, L. Mechanism design and process control of micro EDM for drilling spray holes of diesel injector nozzles. *Precis. Eng.* **2013**, *37*, 213–221.
4. Goda, J.; Mitsui, K. Development of an integrated apparatus of micro-EDM and micro-CMM. *Measurement* **2013**, *46*, 552–562.
5. Vala, M.; Homola, J. Flexible method based on four-beam interference lithography for fabrication of large areas of perfectly periodic plasmonic arrays. *Opt. Express* **2014**, *22*, 18779–18789.
6. Meijer, T.; Beardmore, J.P.; Fabrie, C.G.C.H.M.; Lieshout, J.P.; Notermans, R.P.M.J.W.; Sang, R.T. Structure formation in atom lithography using geometric collimation. *Appl. Phys. B* **2011**, *105*, 703–713.
7. Chou, S.Y.; Krauss, P.R.; Renstrom, P.J. Nanoimprint lithography. *J. Vacuum Sci. Technol. B* **1996**, *14*, 4129–4133.
8. Heckele, M.; Schomburg, W.K. Review on micro molding of thermoplastic polymers. *J. Micromech. Microeng.* **2004**, *14*, R1–R14.
9. Bos, E.J.C. Aspects of tactile probing on the micro scale. *Precis. Eng.* **2011**, *35*, 228–240.
10. Haitjema, H.; Pri, W.O.; Schellekens, P.H.J. Development of a Silicon-based Nanoprobe System for 3-D Measurements. *CIRP Ann. Manuf. Technol.* **2001**, *50*, 365–368.
11. Cui, J.; Li, J.; Feng, K.; Tan, T. Three-dimensional fiber probe based on orthogonal micro focal-length collimation for the measurement of micro parts. *Opt. Express* **2015**, *23*, 26386–26398.
12. Muralikrishnan, V.; Stone, J.; Shakarji, C.; Stoup, J. Performing three-dimensional measurements on micro-scale features using a flexible coordinate measuring machine fiber probe with ellipsoidal tip. *Meas. Sci. Technol.* **2012**, *23*, 025002.

13. Fan1, K.C.; Fei, Y.T.; Yu, X.F.; Chen, Y.J.; Wang, W.L.; Chen, F.; Liu, Y.S. Development of a low-cost micro-CMM for 3D micro/nano measurements. *Meas. Sci. Technol.* **2006**, *17*, 524–532.

14. Schwenke, H.; Waldele, F.; Weiskirch, C.; Kunzmann, H. Opto-tactile Sensor for 2D and 3D Measurement of Small Structures on Coordinate Measuring Machines. *CIRP Ann. Manuf. Technol.* **2001**, *50*, 361–364.

15. Michihata, M.; Hayashi, T.; Adachi, A.; Takaya, Y. Measurement of Probe-stylus Sphere Diameter for Micro-CMM Based on Spectral Fingerprint of Whispering Gallery Modes. *CIRP Ann. Manuf. Technol.* **2014**, *63*, 469–472.

16. Takaya, Y.; Shimizu, H.; Takahashi, S.; Miyoshi, T. Fundamental study on the new probe technique for the nano-CMM based on the laser trapping and Mirau interferometer. *Measurement* **1999**, *25*, 9–18.

17. Masuzawa, T.; Harnasaki, Y.; Fujino, M. Vibroscanning Method for Nondestructive easurement of Small Holes. *CIRP Ann. Manuf. Technol.* **1993**, *42*, 589–592.

18. Kim, B.J.; Masuzawa, T.; Bourouina, T. The vibroscanning method for the measurement of micro-hole profiles. *Meas. Sci. Technol.* **1999**, *10*, 697–705.

19. Weckenmann, A.; Hoffmann, J.; Schuler, A. Development of a tunnelling current sensor for a long-range nano-positioning device. *Meas. Sci. Technol.* **2008**, *19*, 064002.

20. Bauza, M.B.; Hocken, R.J.; Smith, S.T.; Woody, S.C. Development of a virtual probe tip with an application to high aspect ratio microscale features. *Rev. Sci. Instrum.* **2015**, *76*, 095112.

21. Bauza, M.B.; Woody, S.C.; Woody, B.A.; Smith, S.T. Surface profilometry of high aspect ratio features. *Wear* **2011**, *271*, 519–522.

22. Claverley, J.D.; Leach, R.K. Development of a three-dimensional vibrating tactile probe for miniature CMMs. *Precis. Eng.* **2013**, *37*, 491–499.

23. Leach, R.K.; Claverley, J.; Giusca, C.; Jones, C.W.; Nimishakavi, L.; Sun, W.; Tedaldi, M.; Yacoot, A. Advances in engineering nanometrology at the National Physical Laboratory. *Meas. Sci. Technol.* **2012**, *23*, 074002.

24. Murakami, H.; Katsuki, H.; Sajima, T.; Suematsu, T. Study of a vibrating fiber probing system for 3-D micro-structures: Performance improvement. *Meas. Sci. Technol.* **2014**, *25*, 094010.

25. Murakami, H.; Katsuki, A.; Onikura, H.; Sajima, T.; Kawagoshi, N.; Kondo, E. Development of a System for Measuring Micro Hole Accuracy Using an Optical Fiber Probe. *J. Adv. Mech. Des. Syst. Manuf.* **2010**, *4*, 995–1004.

26. Ito, S.; Kodama, I.; Gao, G. Development of a probing system for a micro-coordinate measuring machine by utilizing shear-force detection. *Meas. Sci. Technol.* **2014**, *25*, 064011.

27. Manning, B. *The Use of Non-Contact Thin Gap Sensors in Controlling Coater Gap Uniformity*; Capacitec, Inc.: Ayer, MA, USA, 2012.

28. Furukawa, M.; Gao, W.; Shimizu, H.; Kiyono, S.; Yasutake, M.; Takahashi, K. Slit width measurement of a long precision slot die. *J. Jpn. Soc. Precis. Eng.* **2003**, *69*, 1013–1017.

29. Claverley, J.D.; Leach, R.K. A review of the existing performance verification infrastructure for micro-CMMs. *Precis. Eng.* **2015**, *39*, 1–15.

30. Muralikrishnan, B.; Stone, J.A.; Stoup, J.R. Fiber deflection probe for small hole metrology. *Precis. Eng.* **2006**, *30*, 154–164.

31. Thermo Scientific. Available online: http://www.thermoscientific.com/en/product/ (accessed on 5 May 2016).

32. Ito, S.; Chen, Y.L.; Shimizu, Y.; Kikuchi, H.; Gao, W.; Takahashi, K.; Kanayama, T.; Arakawa, K.; Hayashi, A. Uncertainty analysis of slot die coater gap width measurement by using a shear mode micro-probing system. *Precis. Eng.* **2016**, *43*, 525–529.

33. Working Group 1 of the Joint Committee for Guides in Metrology (JCGM/WG1). JCGM 100. In *Evaluation of Measurement Data—Guide to the Expression of Uncertainty in Measurement (GUM)*; Bureau International des Poids et Mesures: Paris, France, 2008.

Reduction of Liquid Bridge Force for 3D Microstructure Measurements

Hiroshi Murakami, Akio Katsuki, Takao Sajima and Mitsuyoshi Fukuda

Abstract: Recent years have witnessed an increased demand for a method for precise measurement of the microstructures of mechanical microparts, microelectromechanical systems, micromolds, optical devices, microholes, *etc.* This paper presents a measurement system for three-dimensional (3D) microstructures that use an optical fiber probe. This probe consists of a stylus shaft with a diameter of 2.5 µm and a glass ball with a diameter of 5 µm attached to the stylus tip. In this study, the measurement system, placed in a vacuum vessel, is constructed suitably to prevent adhesion of the stylus tip to the measured surface caused by the surface force resulting from the van der Waals force, electrostatic force, and liquid bridge force. First, these surface forces are analyzed with the aim of investigating the causes of adhesion. Subsequently, the effects of pressure inside the vacuum vessel on surface forces are evaluated. As a result, it is found that the surface force is 0.13 µN when the pressure inside the vacuum vessel is 350 Pa. This effect is equivalent to a 60% reduction in the surface force in the atmosphere.

Reprinted from *Appl. Sci.* Cite as: Murakami, H.; Katsuki, A.; Sajima, T.; Fukuda, M. Reduction of Liquid Bridge Force for 3D Microstructure Measurements. *Appl. Sci.* **2016**, *6*, 153.

1. Introduction

Recent years have witnessed an increased demand for a method for precise measurement of the microstructures of mechanical microparts, microelectromechanical systems, micromolds, optical devices, microholes, *etc.* However, precise measurement of the shape of a microstructure with a large length-to-diameter (L/D) ratio is rather difficult because of the difficulty in probe fabrication and sensing methods where the measuring force is very small. Previous works have reported microstructure measurement techniques that employ a variety of probes such as optical probes, vibroscanning probes, vibrating probes, tunneling effect probes, opto-tactile probes, fiber deflection probes, optical trapping probes, and diaphragm probes [1–9].

In a previous study, we developed a system for the measurement of three-dimensional (3D) microstructures using an optical fiber probe that functions as a kind of displacement measuring probe with a small contact force and wide measurement range [10]. In this system, the shaft of the stylus does not need to be rigid for the detection of the measuring force, because the deflection of the stylus is measured by a non-contact method. In general, when the particle size is less than several tens of micrometers, the effects of surface force generated from the van der Waals force, electrostatic force, and liquid bridge force are strengthened and this surface force becomes greater than the force of gravity [11]. We used a fiber stylus with a 5-μm-diameter sphere on its tip, and as a result, we found that the measurement surface draws the stylus tip closer when the tip approaches it and the distance between the stylus tip and the measured surface is less than the regular displacement. When the stylus tip comes into contact with the measured surface, it adheres to the surface and cannot be separated from it. When the measured surface is scanned point by point in the touch-trigger mode (Figure 1a), the measurement time increases because the stylus tip is required to be separated from the measured surface to overcome the surface force. When the measured surface is scanned continuously in the scanning mode (Figure 1b), the measurement accuracy reduces because of the bending of the stylus shaft due to the adhesion. The observed adhesion, which is influenced by environmental factors (e.g., humidity) and the roughness of the measurement surface, is not reproducible. Therefore, in another previous work, we developed a measurement system for 3D microstructures that uses a vibrating fiber probe (Figure 1c) to prevent adhesion of the stylus tip to the surface being measured [12]. In this system, the stylus tip is set to vibrate in a circular motion, where it traces a circle of diameter 0.4 μm in the X-Y plane. However, the stylus tip actually traces an elliptical path. The difference between the profile of the perfect circle and that of the actual elliptical path of the stylus tip leads to measurement error. Moreover, the surface roughness cannot be measured in the scanning mode when using the vibrating fiber probe.

In the present study, the stylus characteristics are examined. Then, the effects of the surface force on the adhesion are analyzed. The results confirm that the primary cause of adhesion is the liquid bridge force. Therefore, the measurement system, placed in a vacuum vessel, is constructed suitably to prevent the adhesion caused by the surface force between the stylus tip and the measured surface. The effects of pressure inside the vacuum vessel on the surface force are evaluated experimentally. The surface force is calculated by assuming that the optical fiber probe is equal to the cantilever of the fixed support. In other words, the surface force is calculated by the deflection of the stylus shaft. There are many techniques for measuring surface forces, such as those involving the surface forces apparatus (SFA), atomic force microscopy (AFM), micro cantilever (MC), optical trapping (OT), *etc.* [13–15]. The measuring method that uses the optical fiber probe is, in principle, similar to the SFA and AFM. As a result, the surface force is found to be 0.13 μN when the pressure inside the vacuum vessel is 350 Pa. This effect is equivalent to a 60% reduction in the surface force in the atmosphere.

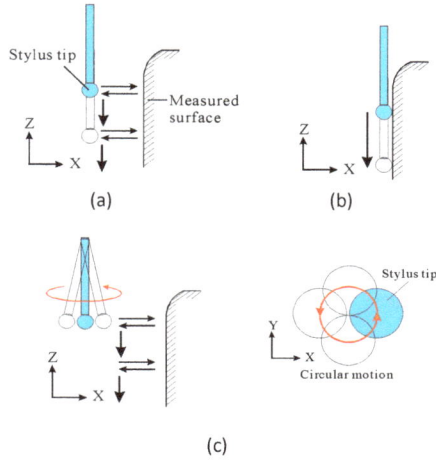

Figure 1. Various types of probes. (**a**) Touch-trigger probe; (**b**) scanning probe; and (**c**) vibrating probe (touch-trigger mode).

2. Measurement Principle

Figure 2 shows a diagram of the developed optical measurement system. The stylus consists of a 2.5-μm-diameter optical fiber to which a 5-μm-diameter glass stylus tip is attached. The total length of the stylus is 0.38 mm. The probing system consists of the fiber stylus, two laser diodes with a 650 nm wavelength (LDX and LDY in the X- and Y-directions, respectively), and two dual-element photodiodes (PX and PY in the X- and Y-directions, respectively). The stylus shaft is fixed to a tube-type piezo driver element in order to perform attitude adjustment of the stylus shaft; the stylus shaft is installed between the laser diodes and the dual-element photodiodes, which are oriented orthogonally. The laser diodes are mounted above the stylus tip, and the focused laser beams irradiate along the X- and Y-directions onto the stylus shaft. The two dual-element photodiodes are located opposite the laser diodes beyond the stylus. The laser beams that pass through the stylus shaft are received by these dual-element photodiodes. The beam intensities detected by the photodiodes are converted into voltages; hereafter, these intensities are denoted as I_{PX1}, I_{PX2}, I_{PY1}, and I_{PY2} (V). The output signal I_X in the X-direction obtained using I_{PY1} and I_{PY2} and the output signal I_Y in the Y-direction obtained using I_{PX1} and I_{PX2} are defined as given in Equations (1) and (2), respectively. A charge-coupled device is employed to monitor the positions of the stylus and test piece during the setting up of the equipment and the measurement.

$$I_X = I_{PY1} - I_{PY2} \tag{1}$$

$$I_Y = I_{PX1} - I_{PX2} \tag{2}$$

141

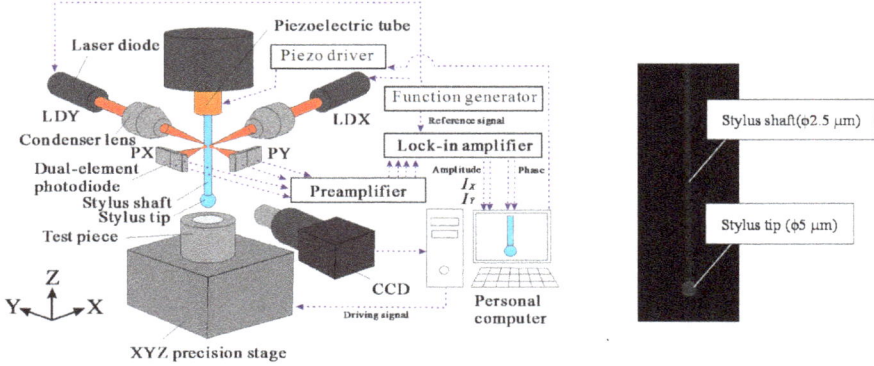

Figure 2. Measurement system using optical fiber probe.

To illustrate the measurement principle of the optical fiber probe, Figure 3 shows a cross-sectional diagram of the X-Y plane of the optical system shown in Figure 2. Before the stylus tip comes into contact with the measured surface, the light intensities measured by each element of the dual-element photodiode are equal (*i.e.*, $I_{PX1} = I_{PX2}$ and $I_{PY1} = I_{PY2}$), as shown in Figure 3a. When the stylus tip comes into contact with the measured surface in the X-direction, the laser-irradiated part of the stylus shaft is displaced, and the light intensities measured by each element of the dual-element photodiode are no longer equal to each other (*i.e.*, $I_{PX1} = I_{PX2}$ and $I_{PY1} > I_{PY2}$), as shown in Figure 3b. When the stylus shaft is displaced in the $+X$-direction, the angle of refraction of the laser beam passing through the stylus shaft in the Y-direction changes owing to a shift in the part of the stylus shaft being irradiated. Additionally, when the stylus tip comes into contact with the measured surface in the Y-direction, the light intensities measured by each element of the dual-element photodiode are no longer equal to each other (*i.e.*, $I_{PX1} < I_{PX2}$ and $I_{PY1} = I_{PY2}$), as shown in Figure 3c. As a result, the contact direction and the magnitude of displacement of the stylus tip can be obtained from the output signals I_X and I_Y. When the stylus tip comes into contact with the measured surface in the Z-direction, the stylus shaft is buckled and deflected. This deflection is also measured using the above-mentioned method.

Because the proposed probe measures the deflection amplitude of the stylus shaft by using a laser-based non-contact method, the stylus shaft does not need to be rigid; this principle also applies to a stylus with a much smaller diameter. This measurement system measures the deflection of the stylus shaft; however, it does not directly measure the displacement of the stylus tip. The noise present in I_X and I_Y is removed via synchronous detection using a lock-in amplifier. The displacement of the stylus is magnified by using it as a rod lens. The surface of the microstructure is

142

measured by recording the displacement of the stylus shaft as well as the coordinates at which the stylus comes into contact with the measured surface.

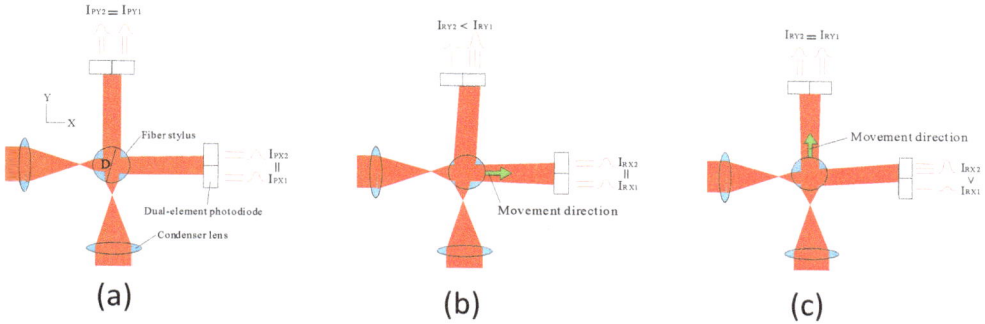

Figure 3. Principle of measurement. (**a**) initial stage; (**b**) displacement in X-direction; and (**c**) displacement in Y-direction.

3. Stylus Characteristics

Figure 4 shows the changes in outputs I_X and I_Y when the stylus tip is displaced in the $\pm X$-direction. Here, the horizontal axis represents the displacement of the stylus tip and the vertical axis represents the changes in I_X and I_Y. It can be seen that when the stylus tip is displaced in the X-direction, barely any change occurs in the output I_Y in the Y-direction and the fiber probe can function as a displacement sensor because the rate of change in I_X can approximate a straight line within a ± 3 μm range in the X-direction.

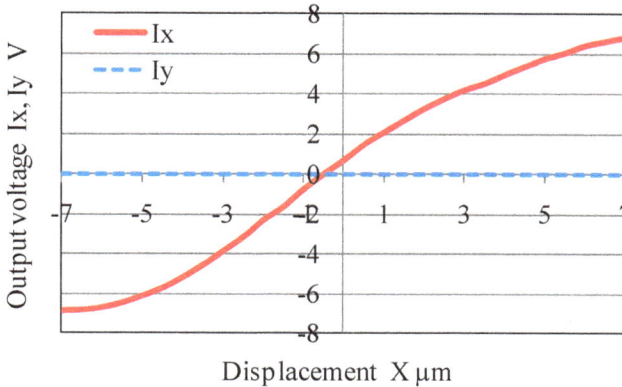

Figure 4. Changes in output voltages I_X and I_Y induced by displacement in $\pm X$-direction.

143

4. Effects of Surface Force

4.1. *Analyses of Liquid Bridge Force, Van der Waals force, and Electrostatic Force*

In general, when the particle size is less than several tens of micrometers, the effects of the surface force generated from the van der Waals force, electrostatic force, and liquid bridge force are strengthened and this surface force becomes greater than the force of gravity [11]. Figure 5 shows a schematic diagram of the surface force between the stylus tip and the measurement surface. Because the fiber stylus has a 5-μm-diameter sphere on its tip, the measurement surface draws the stylus tip closer when the stylus tip approaches it, and the distance between the stylus tip and the measured surface is less than the regular displacement. When the stylus tip comes into contact with the measured surface, it adheres to the surface and cannot be separated from it. When the measured surface is scanned point by point in the touch-trigger mode, the measurement time increases because the stylus tip is required to be separated from the measured surface on account of the surface force. When the measured surface is scanned continuously in the scanning mode, measurement accuracy reduced because of the bending of the stylus shaft due to the adhesion. The observed adhesion, which is influenced by environmental factors (e.g., humidity) and the roughness of the measured surface, is not reproducible.

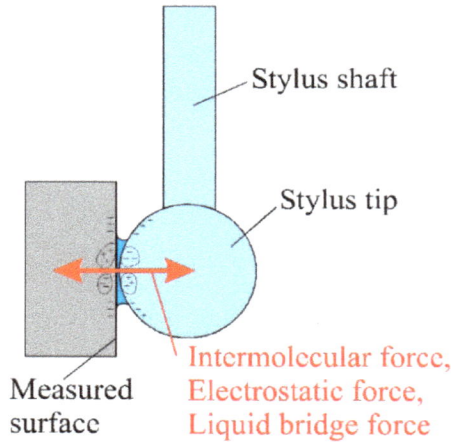

Figure 5. Schematic diagram of surface force between stylus tip and measured surface.

These surface forces are analyzed with the aim of investigating the causes of adhesion. First, we examine the liquid bridge force. This force has a large effect on adhesion in a humid environment; it is generated by the capillary action of condensed water near the contact zone of the stylus tip and measured surface. This force increases with increasing relative humidity. The liquid bridge force F_1

144

experienced in the case of a stylus tip of diameter d and under a relative humidity of %RH can be calculated using the empirical formula in Equation (3) [11]. This is the empirical formula of the liquid bridge force between a particle and a plate with a hard and clean surface. As indicated by this formula, the liquid bridge force increases with increasing relative humidity.

$$F_1 = 0.15d\{0.5 + 0.0045 \times (\%RH)\} \tag{3}$$

Next, we examine the van der Waals force. This force acts between molecules, and it is a general force of interaction that acts between objects at all times. The van der Waals force F_V can be calculated using the Hamaker constant A and the distance z between the stylus tip and the measured surface as given in Equation (4) [11]. The Hamaker constant A is a constant specific to materials, and it is expressed as in Equation (5), where n and Λ are the number of atoms per unit volume and the constant of proportionality of London forces, respectively. When the molecules of the stylus tip and the measured surface come close to each other by the van der Waals force, intermolecular repulsive force is generated between the molecules of the stylus tip and the measured surface. The molecules are stabilized at the most stable position, which corresponds to the position with the smallest potential energy. At this instant, the separation distance z is about 0.4 nm.

$$F_V = Ad/(12z^2) \tag{4}$$

$$A = n^2 \pi^2 \Lambda \tag{5}$$

Finally, the electrostatic force generated by an interaction between a charged particle and an uncharged surface is examined. When a charged particle comes into contact with an uncharged surface, the generated electrostatic force F_e can be calculated using Equation (6) in the case that the distance between the stylus tip and the measured surface is negligible because it is remarkably small compared with the diameter of the stylus tip [11].

$$F_e = \frac{\pi}{4\varepsilon_0} \cdot \frac{\varepsilon - \varepsilon_0}{\varepsilon + \varepsilon_0} \cdot d^2 \sigma^2 \tag{6}$$

where ε_0, ε, and σ are the dielectric constant of the vacuum, the dielectric constant of the measured surface, and the marginal surface density of the charge of the stylus tip, respectively.

Figure 6 shows the relationship between the stylus tip diameter and the surface forces calculated using Equations (3), (4), and (6). Here, the horizontal axis represents the stylus tip diameter and the vertical axis represents the van der Waals force, electrostatic force, liquid bridge force, and gravity. It can be seen from this figure

145

that when the stylus tip diameter is less than about 1 mm, the effects of surface force generated from the van der Waals force, electrostatic force, and liquid bridge force are strengthened and the surface force becomes greater than the force of gravity.

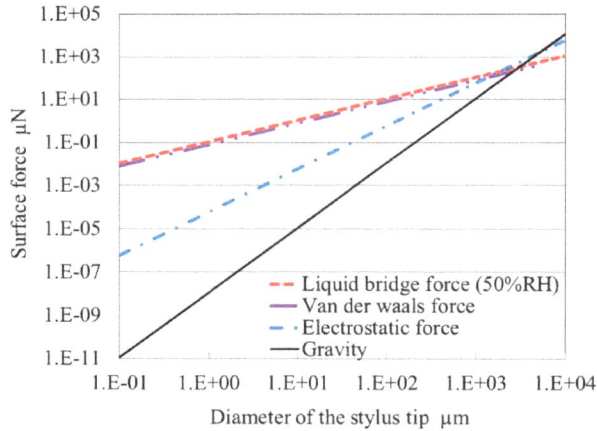

Figure 6.   Relationship between stylus tip diameter and surface forces: $A\,(SiO_2) = 1.5 \times 10^{-20}$ J, $z = 0.4$ nm, $\varepsilon_0 = 8.85 \times 10^{-12}$ F·m^{-1}, $\varepsilon = \infty$, $\sigma = 26.5\ \mu C \cdot m^{-2}$.

This confirms that the primary cause of adhesion is the liquid bridge force. The van der Waals force and liquid bridge force are approximately equal. However, the electrostatic force is much smaller than the van der Waals force and liquid bridge force. When the surface forces are calculated using Equations (3), (4), and (6) for a stylus tip 5 μm in diameter, the liquid bridge force (50% RH), van der Waals force, and electrostatic force are 0.54 μN, 0.36 μN, and 0.0016 μN, respectively. In this case, the sum of the surface forces is approximately 0.9 μN. Figure 7 shows the relationship between the relative humidity and the sum of surface forces ($F_l + F_v + F_e$) calculated using Equations (3), (4), and (6). It can be seen from this figure that the liquid bridge force increases with increasing relative humidity.

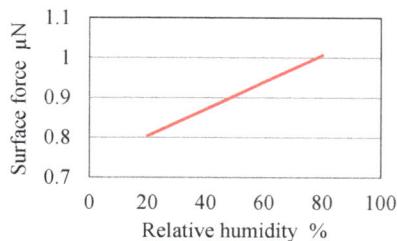

Figure 7. Relationship between relative humidity and surface force (stylus tip diameter: 5 μm).

146

4 2. Effect of Relative Humidity on the Liquid Bridge Force

As stated in the discussion above, the primary cause of adhesion is the liquid bridge force, which increases with increasing relative humidity. When the measured surface is scanned point by point in the touch-trigger mode, the measurement time increases because the stylus tip is required to be separated from the measured surface on account of the surface force. When the measured surface is scanned continuously in the scanning mode, the measurement accuracy reduces owing to the bending of the stylus shaft due to the adhesion. Therefore, in our previous work, a vibration mechanism was introduced to prevent the adhesion of the stylus tip to the measurement surface caused by the surface force [12]. By this mechanism, the stylus tip is set to vibrate in a circular motion, where it traces a circle of diameter 0.4 μm in the X-Y plane. However, the stylus shaft actually traces an elliptical path. The difference between the profile of the perfect circle and that of the actual elliptical path of the stylus tip leads to measurement errors. Moreover, the surface roughness cannot be measured in the scanning mode when using the vibrating fiber probe.

Therefore, the measurement system, placed in a vacuum vessel, is constructed suitably to prevent adhesion of the stylus tip to the measured surface caused by the surface force. Figure 8 shows a photograph of the measurement system in the vacuum vessel. The effects of pressure inside the vacuum vessel on the surface force are evaluated experimentally. In the experiment, the relative humidity is 61% and the temperature is 25.1 °C.

First, the measurement method of the surface force is explained as follows. There are many techniques for measuring surface forces, such as those involving the surface forces apparatus (SFA), atomic force microscopy (AFM), micro cantilever (MC), optical trapping (OT), etc. [13–15]. The SFA can directly measure the force between two surfaces in controlled vapors or immersed in liquids. The force sensitivity of the SFA is about 10 nN. The SFA contains two curved molecularly smooth surfaces of mica between which the interaction forces are measured using a variety of force-measuring springs. AFM is, in principle, similar to the SFA except that forces are measured not between two macroscopic surfaces but between a fine tip and a surface. The force sensitivity of AFM is 1–10 pN. The measuring method using the optical fiber probe is, in principle, similar to the SFA and AFM. As shown in Figure 9, after the stylus tip comes into contact with the measured surface, the stylus tip is displaced in the backward direction by using the piezoelectric stage, due to which it separates from the measured surface. Because the stylus tip is displaced while maintaining its contact with the measured surface, the stylus shaft is deflected and the force required to separate the stylus tip from the measured surface is generated by the deflection of the stylus shaft. With an increase in the displacement of the stylus tip, this required force also increases. Therefore, the stylus tip can be separated from the measured surface when the displacement D of the stylus tip exceeds a certain value.

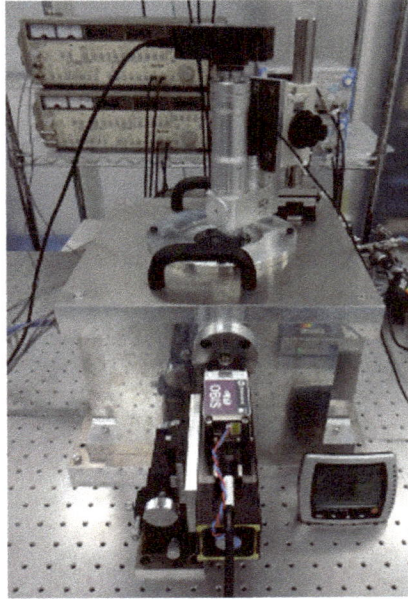

Figure 8. Photograph of measurement system in vacuum vessel.

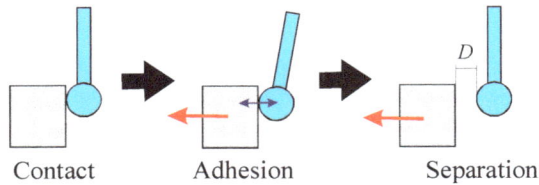

| Contact | Adhesion | Separation |

Figure 9. Experimental apparatus for measurement of surface force.

The load acting on the stylus tip, *i.e.*, the surface force, is calculated by assuming that the optical fiber probe, 2.5 μm in diameter and 0.38 mm in length, is equal to the cantilever of the fixed support. In other words, the surface force is calculated by the deflection of the stylus shaft. The surface force F is calculated using Equation (7).

$$F = \frac{3EI}{L^3} \cdot D \tag{7}$$

where E, I, and L are the Young's modulus, the geometrical moment of inertia, and the length of the stylus shaft, respectively. Figure 10 shows the load acting on the stylus tip under the assumption that the stylus shaft (Young's modulus $E = 72$ GPa) is equivalent to the cantilever of the fixed support. In this figure, the horizontal axis represents displacement of the stylus tip, and the vertical axis represents the load acting on the stylus tip. For example, when the displacement D required to separate the stylus tip from the measured surface is about 20 μm, the load acting on the stylus tip is about 0.15 μN immediately before the stylus tip is separated from the measured surface as shown in Figure 10. Therefore, this value of 0.15 μN is regarded as the surface force. In this experiment, the spring constant of the stylus shaft is not calibrated. However, to accurately measure the surface force, it is necessary to calibrate it [16–19].

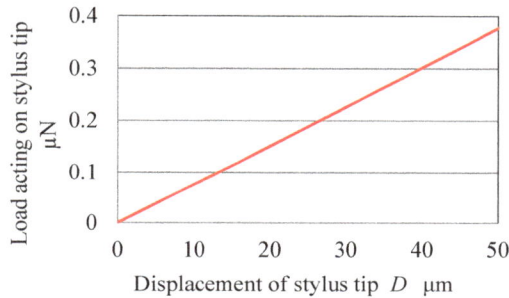

Figure 10. Load acting on the stylus tip (stem length: 0.38 mm, stem diameter: 2.5 μm, and ball diameter: 5 μm).

Figure 11 shows the relationship between the pressure inside the vacuum vessel and the relative humidity. Further, Figure 12 shows the relationship between the pressure inside the vacuum vessel and the surface forces. Finally, Figure 13 shows the relationship between the relative humidity and the surface force. From Figure 13, it is confirmed that surface force decreases with decreasing relative humidity. It is found that the surface force is 0.13 μN when pressure inside the vacuum vessel and relative humidity are 350 Pa and 7% *RH*, respectively. This effect is equivalent to a 60% reduction in the surface force in the atmosphere. However, a large residual surface force still acts between the stylus tip and the measured surface. This is because the amount of van der Waals force is almost the same as that of the liquid bridge force. Therefore, the surface force is thought to reduce by slightly a little more than 60%. When the stylus tip diameter is 5 μm as shown in Figure 6, the theoretical van der Waals force in a vacuum environment is about 0.36 μN. However, according to the obtained experimental results, the van der Waals force in a vacuum environment is about 0.13 μN. Thus, the theoretical and experimental results are different. Because the amount of van der Waals force is affected by the surface roughness of the stylus tip and that of the measurement surface [11], the surface roughness of the stylus tip and that of the measurement surface are, in turn, thought to be affected by the amount of van der Waals force.

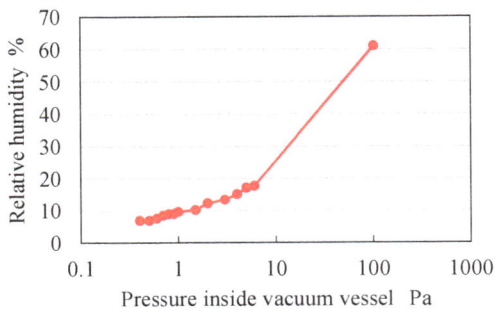

Figure 11. Relationship between pressure inside vacuum vessel and relative humidity.

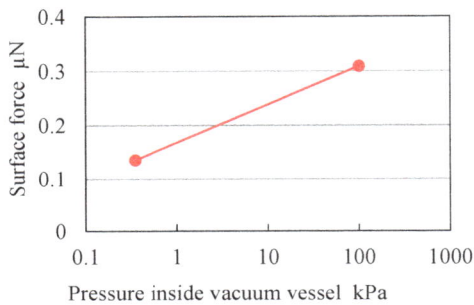

Figure 12. Relationship between pressure inside vacuum vessel and surface forces.

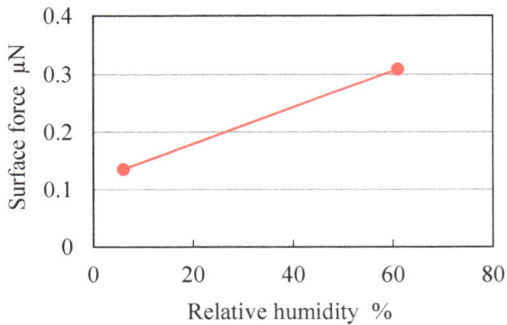

Figure 13. Relationship between relative humidity and surface forces.

According to the above-mentioned experimental results, it is possible to reduce the liquid bridge force considerably by using a vacuum vessel. However, in scenarios where the use of a vacuum vessel is difficult, the stylus tip coated with a water-repellent coating has the potential to reduce the liquid bridge force. In order to reduce the van der Waals force, a stylus tip made of a material with a small Hamarker constant may be used. For example, the Hamaker constants of SiO_2, Fe, Si, Cu, and Au are 8.55, 21.2, 25.6, 28.4, and 45.5×10^{-20} J, respectively. However, in practice, the use of a material with a small low Hamarker constant can be difficult. In such cases, a stylus tip coated with a material with a small Hamarker constant may be used. The effect of the material of the stylus tip can be made negligible by using a coating that is several nanometers thick. The influence of the electrostatic force on the surface force is small; however, it is believed that various methods of removal of electricity or antistatic agents could be useful in reducing the electrostatic force.

5. Conclusions

In this study, the stylus characteristics are examined, and then, surface forces are analyzed in order to investigate the causes of adhesion, which reduces the measurement accuracy. The analysis results of surface forces reveal the liquid bridge force to be the primary cause of adhesion. Therefore, the measurement system placed in the vacuum vessel is constructed suitably to prevent adhesion of the stylus tip to the measured surface because of the surface forces, mainly by reducing the liquid bridge force. As a result, it is found that the surface force is 0.13 μN when the pressure inside the vacuum vessel is 350 Pa. This effect is equivalent to a 60% reduction in the surface force in the atmosphere.

Acknowledgments: This study was partly supported by a research grant from the Mitutoyo Association for Science and Technology and by JSPS KAKENHI Grant Number 26420392.

Author Contributions: Hiroshi Murakami conceived and wrote the paper; Akio Katsuki and Takao Sajima manufactured the vacuum vessel and analyzed the data; Hiroshi Murakami and Mitsuyoshi Fukuda performed the experiments and analyzed the surface forces.

Conflicts of Interest: The authors declare no conflict of interest.

References

1. Masuzawa, T.; Hamasaki, Y.; Fujino, M. Vibroscanning method for nondestructive measurement of small holes. *CIRP Ann.* **1993**, *42*, 589–592.
2. Hidaka, K.; Saito, A.; Koga, S. Study of a micro-roughness probe with ultrasonic sensor. *CIRP Ann.* **2008**, *57*, 489–492.
3. Bauza, M.B.; Hocken, R.J.; Smith, S.T.; Woody, S.C. Development of a virtual probe tip with an application to high aspect ratio microscale features. *Rev. Sci. Instrum.* **2005**, *76*.

4.   Claverley, J.D.; Leach, R.K. A vibrating micro-scale CMM probe for measuring high aspect ratio structures. *Microsyst. Technol.* **2010**, *16*, 1507–1512.

5.   Shiramatsu, T.; Kitano, K.; Kawata, M.; Mitsui, K. Development of a measuring method for shape and dimension of micro-components. Modification to the original measuring system, calibration of the probes and the results of dimensional measurements. *JSME Int. J. Ser. C Mech. Syst.* **2002**, *68*, 267–274.

6.   Schwenke, H.; Wäldele, F.; Weiskrich, C.; Kunzmann, H. Opto-Tactile sensor for 2D and 3D measurement of small structures on coordinate measuring machines. *CIRP Ann.* **2001**, *50*, 361–364.

7.   Muralikrishnan, B.; Stone, J.A.; Stoup, J.R. Fiber deflection probe for small hole metrology. *Precis. Eng.* **2006**, *30*, 154–164.

8.   Michihata, M.; Takaya, Y.; Hayashi, T. Development of the nano-probe system based on the laser-trapping technique. *CIRP Ann.* **2008**, *57*, 493–496.

9.   Liebrich, T.; Knapp, W. New concept of a 3D-probing system for micro-components. *CIRP Ann.* **2010**, *59*, 513–516.

10. Murakami, H.; Katsuki, A.; Onikura, H.; Sajima, T.; Kawagoishi, N.; Kondo, E. Development of a system for measuring micro hole accuracy using an optical fiber probe. *J. Adv. Mech. Des. Syst. Manuf.* **2010**, *5*, 995–1004.

11. Okuyama, K.; Masuda, H.; Morooka, S. *Biryushi Kogaku*; Ohmsha: Tokyo, Japan, 1992.

12. Murakami, H.; Katsuki, A.; Sajima, T.; Suematsu, T. Study of a vibrating fiber probing system for 3-D micro-structures: Performance improvement. *Meas. Sci. Technol.* **2014**, *25*

13. Israelachvili, J.N. *Intermolecular and Surface Forces: Revised Third Edition*; Academic Press: Waltham, MA, USA, 2011; pp. 227–247.

14. Lee, M.; Kim, B.; Kim, J.; Jhe, W. Noncontact friction via capillary shear interaction at nanoscale. *Nat. Commun.* **2015**, *6*.

15. Yongho, S.; Wonho, J. Atomic force microscopy and spectroscopy. *Rep. Prog. Phys.* **2008**, *71*.

16. Guebum, H.; Ahn, H.S. Calibration of effective spring constants of colloidal probes using reference cantilever method. *Coll. Surf. A Physicochem. Eng. Asp.* **2015**, *489*, 86–94.

17. Song, Y.P.; Wu, S.; Xu, L.Y.; Zhang, J.M.; Dorantes-Gonzalez, D.J.; Fu, X.; Hu, X.D. Calibration of the effective spring constant of ultra-short cantilevers for a high-speed atomic force microscope. *Meas. Sci. Technol.* **2015**, *26*.

18. Slattery, A.D.; Blanch, A.J.; Ejov, V.; Quinton, J.S.; Gibson, C.T. Spring constant calibration techniques for next-generation fast-scanning atomic force microscope cantilevers. *Nanotechnology* **2014**, *25*.

19. Slattery, A.D.; Blanch, A.J.; Quinton, J.S.; Gibson, C.T. Calibration of atomic force microscope cantilevers using standard and inverted static methods assisted by FIB-milled spatial markers. *Nanotechnology* **2013**, *24*.

Development of a High-Precision Touch-Trigger Probe Using a Single Sensor

Rui-Jun Li, Meng Xiang, Ya-Xiong He, Kuang-Chao Fan, Zhen-Ying Cheng, Qiang-Xian Huang and Bin Zhou

Abstract: To measure various components with nano-scale precision, a new high-precision touch-trigger probe using a single low-cost sensor for a micro-coordinate measuring machine (CMM) is presented in this paper. The sensor is composed of a laser diode, a plane mirror, a focusing lens, and a quadrant photo detector (QPD). The laser beam from the laser diode with an incident angle is reflected by the plane mirror and then projected onto the quadrant photo detector (QPD) via the focusing lens. The plane mirror is adhered to the upper surface of the floating plate supported by an elastic mechanism, which can transfer the displacement of the stylus's ball tip in 3D to the plane mirror's vertical and tilt movement. Both motions of the plane mirror can be detected by respective QPDs. The probe mechanism was analyzed, and its structural parameters that conform to the principle of uniform sensitivity and uniform stiffness were obtained. The simulation result showed that the stiffness was equal in 3D and less than 1 mN/μm. Some experiments were performed to investigate the probe's characteristics. It was found that the probe could detect the trigger point with uniform sensitivity, a resolution of less than 5 nm, and a repeatability of less than 4 nm. It can be used as a touch-trigger probe on a micro/nano-CMM.

Reprinted from *Appl. Sci.* Cite as: Li, R.-J.; Xiang, M.; He, Y.-X.; Fan, K.-C.; Cheng, Z.-Y.; Huang, Q.-X.; Zhou, B. Development of a High-Precision Touch-Trigger Probe Using a Single Sensor. *Appl. Sci.* **2016**, *6*, 86.

1. Introduction

With the development of various micro-fabrication technologies, many miniaturized structures and components with nano-scale precision have been produced in recent twenty years. Accordingly, many micro-/nano-coordinate measuring machines (micro-/nano-CMMs) have been proposed to satisfy the urgent demand for the dimensional measurement of micro parts [1–4]. Many touch probing systems that can be equipped onto micro-/nano-CMMs have also been developed, such as (a) the capacitive probe that uses at least three high-precision capacitive sensors to detect the arm's displacement of the floating plate of the probe [5,6], (b) the strain gauge probe that adheres strain gauges on the membrane or cantilevers symmetrically to detect the ball tip's motion using the piezo-resistive effect [7–10], (c) the inductive probe that uses three high-precision inductive sensors and a complicated flexure hinges to construct the probe head [11], (d) the fiber probe that uses the imaging system to detect the ball tip's motion [12–14] or uses long Bragg gratings to detect the axial motion of the probe tip [15] or its 3D motion [16,17], and (e) the optical sensing probe that uses different optical principles to detect the probe motion, such as position detector, focus sensor, interferometer, auto-collimator, *etc.* [18–26]. Although all the above probes have good sensitivity, accuracy, resolution, repeatability, and stiffness, their permissible measurement ranges are limited since their adopted sensors do not tolerate a large tilt or a large translation of the probe tip. In addition, due to the use of multiple sensors for detecting 3D motions of the probe, their costs are still high.

The QPD (quadrant photo detector)-based angle sensor has high accuracy and high sensitivity, which has been verified in some probes [20,25,26], especially in Atomic Force Microscope (AFM) [23]. A new touch-trigger probe with a simple structure, small size, and low cost is proposed in this paper. Only one QPD-based two-dimensional angle sensor is used in this probe, which can simultaneously detect the trigger signals of the probe tip in 3D. Targets of the probe's design include: (1) that the permissible range of the probe is more than ±6 µm; (2) that the probe has equal sensitivity and equal stiffness in three dimensions; (3) that the stiffness is less than 1 mN/µm; (4) that the repeatability of trigger measurement is less than 5 nm ($K = 2$); and (5) that the cross-sectional diameter is less than 40 mm. The design principle and optimal parameters of the innovative touch-trigger probe are addressed in this paper. Experimental results also show the characteristics of the probe.

2. Structure and Principle

This touch trigger probe, shown in Figure 1, mainly consists of three components: a QPD-based sensor, a floating plate, and a tungsten stylus with a ruby ball tip. A plane mirror is adhered to the upper surface of the floating plate. The stylus is mounted on the center position on the lower surface of the floating plate. Four V-shaped leaf springs, fixed to the probe housing, are designed to connect the floating plate. When a contact force is applied to the ball tip of the stylus, the floating plate and the stylus perform a rigid body motion. Simultaneously, the four leaf springs yield corresponding elastic deformations by the floating plate, and the plane mirror on the floating plate tilts along horizontal axes or displaced in the vertical axis. The probe housing is made of an aluminum cylinder. The QPD, the focus lens, and the circuit board are imbedded in the probe. A laser diode is placed on an adjusting mechanism whose elastic component is formed by folding a beryllium–copper alloy sheet. The adjusting mechanism is applied to assure the reflected laser beam being focused onto the centre of the QPD in the initial state.

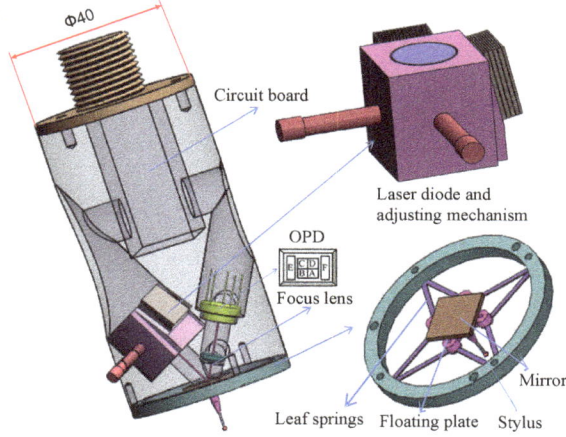

Figure 1. The sketch map of the probe.

Figure 2 shows the optical path of the sensor system, which is composed of a laser diode, a plane mirror, a focusing lens, and a QPD. The laser beam from the laser diode with an incident angle is reflected by the plane mirror and then projected onto the QPD via the focusing lens. A tilt angle or a vertical displacement of the plane mirror causes a lateral shift of the focused light spot on the QPD. The QPD outputs the light energy of each quadrant photo detector into an electrical current signal. By applying an appropriate resistance to these current signals, four output voltage signals (V_A, V_B, V_C, and V_D) can be obtained. Two-dimensional shifts of the focused light spot caused by the motion of the plane mirror can be expressed by Equations (1) and (2), in which k_1 and k_2 are constants [26]. When the ball tip is contacted in any direction by the workpiece, the probe generates a trigger signal at the same time.

$$x = k_1 \left[(V_A + V_D) - (V_B + V_C) \right] \tag{1}$$

$$y = k_2 \left[(V_A + V_B) - (V_C + V_D) \right] \tag{2}$$

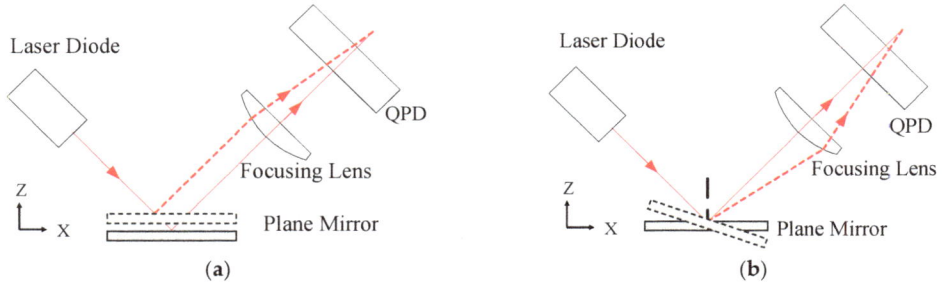

Figure 2. Optical paths of the sensor. (**a**) Mirror in vertical displacement and (**b**) mirror in tilt motion.

3. Analysis and Design of the Probe

3.1. The Sensitivity Analysis

The principle of detecting the Z-motion of the probe, when the ball tip is touched along this direction, is shown in Figure 3. The optical beam with an incidence angle of α is reflected by the plane mirror and focused onto the QPD. The reflected beam is shifted from point D to point E on the focusing lens, and the focused spot is shifted from point B to point C on the QPD if the QPD is placed in front of the focal point of the lens. From the geometrical relationship in Figure 3, we have $\sin \alpha = (\overline{FG}/\overline{FH})$ and $\overline{FG} = \delta_v$ (δ_v is the vertical displacement of the probe tip). The line of FI is perpendicular to the line of AH so that

$$\sin\angle FHI = \sin(\pi - 2\alpha) = sin\,(2\alpha) = \overline{FI}/\overline{FH} \tag{3}$$

Therefore,

$$\overline{DE} = \overline{FI} = 2 \cdot \delta_v \cdot \cos \alpha \tag{4}$$

Because $(\overline{AD}/\overline{AB}) = (\overline{DE}/\overline{BC})$, and $\overline{AD}$ (the focal length of the focusing lens) and $\overline{BD}$ are donated by f and m, respectively, we have

$$\overline{BC} = 2 \cdot \delta_v \cdot (1 - m/f) \cdot \cos \alpha \tag{5}$$

The output voltage (U_v) from QPD is proportional to the displacement of the focused spot ($\overline{BC}$), namely, $U_v = k \cdot \overline{BC}$ and k is a constant. Therefore, the sensitivity in the vertical direction can be obtained as:

$$S_v = U_v/S_v = 2k \cdot (1 - m/f) \cdot \cos \alpha \tag{6}$$

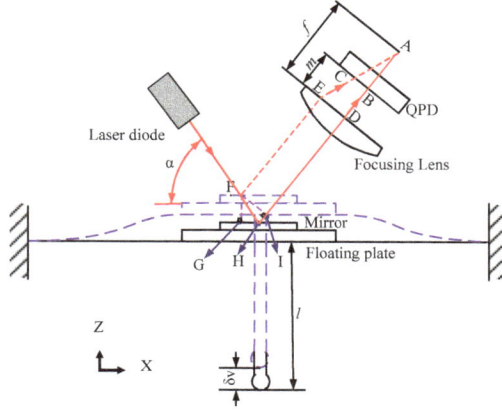

Figure 3. Detecting principle of the probe in Z-motion.

In the same way, the principle of the detecting probe in a horizontal tilt motion, when the probe tip is touched in the horizontal direction and results in a δ_h displacement is shown in Figure 4. The rotated angle and the length of the stylus are denoted as θ and l, respectively. From the geometrical relationship of the optical path, we have $\overline{MN} \approx \overline{DE}$ and $\overline{ON} \approx \overline{OD} = n$. Because, in practice, θ is very small, it is assumed $\sin \theta = (\delta_h/l) \approx \theta$. The stylus and the floating plate can be regarded as a rigid body and $\angle NOM = 2\theta$, so that

$$\tan 2\theta = (\overline{MN}/\overline{ON}) = (\overline{DE}/n) \approx 2\theta \tag{7}$$

$$\overline{DE} = 2n\delta_h/l \tag{8}$$

Because $\overline{AB}/\overline{AD} = \overline{BC}/\overline{DE} = (f - m)/f$ and $U_h = k \cdot \overline{BC}$, the horizontal sensitivity can be obtained:

$$S_h = U_h/\delta_h = (2nk/l) \cdot (1 - m/f) \tag{9}$$

In order to have the characteristic of uniform sensitivity, we can let Equation (6) be equal to Equation (9), yielding:

$$\cos \alpha = n/l \tag{10}$$

157

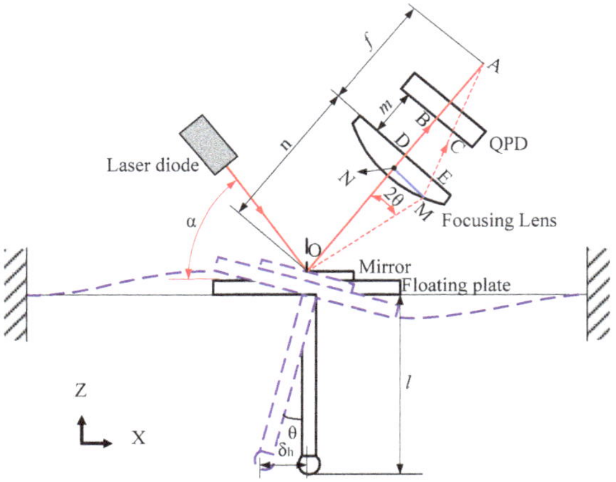

Figure 4. Detecting principle of the probe in horizontal tilt motion.

3.2. The Stiffness Analysis

The free-body diagram of the floating plate is illustrated in Figure 5. T, P, M, F are torque, shear force, bending moment, and contact force, respectively. The characteristics of this probe between the contact force and the ball tip's motion were analyzed, and the stiffness models in horizontal and vertical directions are as follows [26,27].

$$K_y = F_y/\delta_{b,y} = \frac{4}{L^3 l^2} \left[GJL^2 + 2EI \left(6r^2 + 3r \sin \beta + 3rL \sin \beta + L\right) + 2EI \left(6a^2 + 3a \cos \beta + 3aL \cos \beta + L\right) \right] \quad (11)$$

$$K_Z = \frac{F_Z}{\delta_{b,Z}} = \frac{96EI}{L^3} \quad (12)$$

where $I = \dfrac{wt^3}{12}$, $\mathbf{J} = \dfrac{wt^3}{16}\left(\dfrac{16}{3} - 3.36\dfrac{t}{w}\right)$, and $G = \dfrac{E}{2(1+v)}$; t, w, and L are the thickness, width, and length of the leaf springs, respectively; E and v are Young's modulus and Poisson's ratio of the leaf springs, respectively; l is the length of the stylus; r is the arm length of the floating plate; and a is one-half of the arm width.

3.3. Optimal Design for the Probe

The optimal parameters of the probe (see Table 1) have been evaluated according to the constrained conditions that the limited stiffness should be less than 1 mN/μm, i.e., $K_v = K_h \leqslant 1$ mN/μm, and the maximum cross-sectional diameter of the probe head should be smaller than 40 mm.

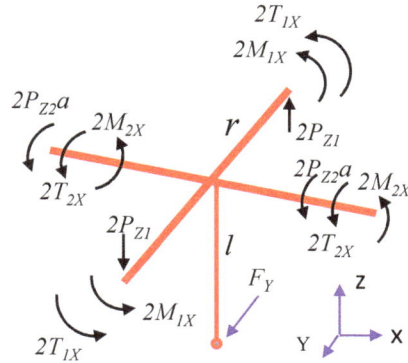

Figure 5. Free-body diagram of the floating plate.

Table 1. The parameters of the probe.

Parameter	Value
Material of the leaf springs	beryllium-copper alloy
Young's modulus of the leaf springs (GPa)	130
Leaf thickness × width × length (mm)	0.1 × 2 × 13
Material of the floating plate	aluminum alloy
Young's modulus of the floating plate (GPa)	71
floating plate arm thickness × width × length (mm)	1.5 × 2 × 5.5
Weight of the floating plate (g)	0.2
Material of the stylus	tungsten stylus with a ruby ball tip
Young's modulus of the stylus (GPa)	193
Length of the stylus (mm)	10
Diameter of the ball tip (mm)	0.5

Moreover, based on the constrained conditions of uniform sensitivity (Equation 10) and the measurement range being at least ±6 μm, the other optimal parameters have also been obtained as: $\alpha = 53°$, $f = 9.8$ mm, $n = 6$ mm, and $m = 4.9$ mm.

A finite element analysis for the probe was also performed using ANSYS V15 software (Ansys Inc, Canonsburg, PA, USA, 2014) so as to verify the stiffness model. The parameters listed in Table 1 were used. Structural deformations are shown in Figure 6a,b when 1 mN of force is applied in the vertical and horizontal directions, respectively. The resulting tip displacement of nearly 1 μm is almost the same in both directions, corresponding to a uniform stiffness of 0.926 μm. Figure 6c illustrates the uniformity of the probe's stiffness along the X-Y plane. In addition, the experimental verification for the stiffness of the floating mechanism was also conducted using a high sensitivity force sensor [26]. The stiffness measurements were 0.954 N/mm in the Z-direction and 0.927 N/mm in the horizontal direction, which is quite consistent with the finite element analysis and analytical results [27].

159

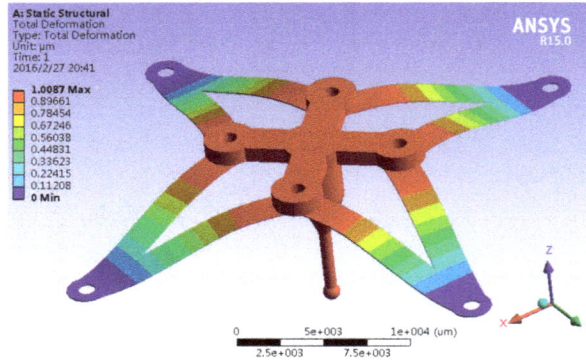

(a)

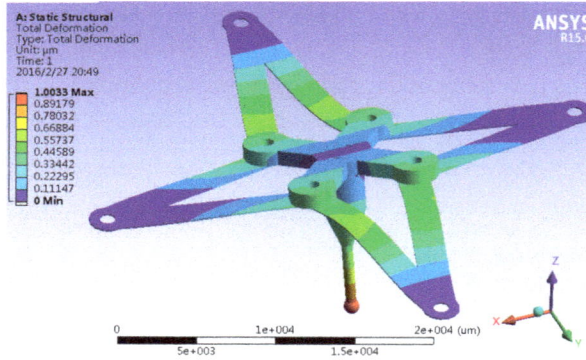

(b)

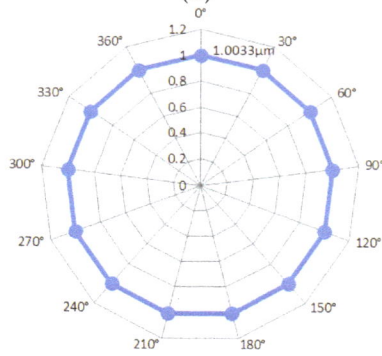

(c)

Figure 6. The tip's displacements when being touched by 1 mN force. (**a**) Touched in Z direction; (**b**) touched in Y direction and (**c**) touched with a different angle in XY plane.

4. Experimental Results and Discussion

In order to test the performance of the designed probe, an experimental setup was carried out. As shown in Figure 7, a stable stand frame was used to hold the probe. Four high accuracy gauge blocks were used to form a 2 mm × 2 mm square hole, which was able to contact the ball tip from different horizontal directions. A ball tip of 1 mm diameter was adopted. The clamped set of gauge blocks were mounted on a high-precision 3D nano-positioning stage made by Physik Instrumente (PI, model P-561.3CD with a repeatability of 2 nm and a distance of travel of 100 μm in each direction, Physik Instrumente Co. Ltd., Karlsruhe, Germany). A normal 2D high-precision stage was used to manually adjust the initial position of the square hole. A Data Acquisition (DAQ) card from National Instruments (PCI-6251, National Instruments Co. Ltd., Austin, TX, USA) was use to recorded the voltage signals from the angle sensor.

Figure 7. Photo of the experimental setup.

4.1. Probe Drift

In order to reduce the influence of the environmental temperature fluctuation, a low-cost vibration-free constant-temperature mini chamber was developed by our group [28,29]. The experimental setup, as shown in Figure 7, was put in the chamber, and the probe was adjusted to output a non-zero value. The stability of the probe was not investigated until the temperature in the chamber had been stabilized to within 20 ± 0.05 °C, as shown in Figure 8. Figure 9 shows the stability test results of the probe. It can be seen that the total drift of the probe was less than 5 nm for duration of three hours, and the fluctuation was less than 3 nm after two hours. The stability of the probe is thus confirmed.

161

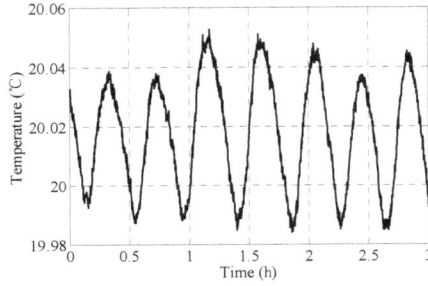

Figure 8. The temperature in the chamber.

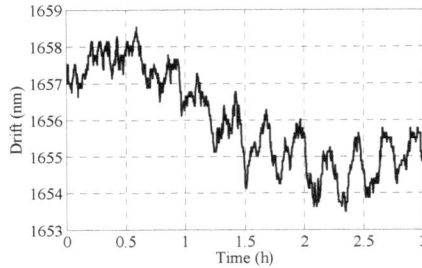

Figure 9. Stability test results of the probe.

4.2. Probe Repeatability

The touch-trigger repeatability of the probe along $X+$, $X-$, $Y+$, $Y-$, and Z directions was also tested. The probe was touched seven times in each direction with a displacement of 1 μm of the Physik Instrumente (PI) stage after the temperature in the chamber had stabilized. The residual errors of the test results in each direction are shown in Table 2. The maximum single directional repeatability was 4 nm ($K = 2$).

Table 2. Results of the trigger test.

Item	Residual Errors (nm)				
	$X+$	$X-$	$Y+$	$Y-$	Z
1st	−0.6	−1.1	−2.7	−1.4	−1.9
2nd	2.9	−0.8	0.2	1.3	−1.4
3rd	0.6	1.9	−1.8	−0.3	−1.2
4th	−2.2	0.7	1.6	1.3	0.1
5th	0.6	−2.6	2.1	0.7	−2.4
6th	−3.2	0.8	0.1	−0.3	−2.8
7th	1.3	2.0	1.2	−1.0	−1.0
standard deviation	2.0	1.7	1.8	1.1	1.0
repeatability ($K = 2$)	4.0	3.4	3.6	2.2	2.0

4 3. Sensitivity and Permissible Range

Figure 10 shows the sensitivity and permissible range of the probe. It can be found that the sensitivities of the probe in horizontal and vertical directions were nearly equal and coincident with the design target quite well when the ball tip's displacement was within 1 μm. In other words, the probe has a uniform sensitivity within a trigger range of 1 μm that is large enough for the touch-trigger measurement. Figure 10 also illustrates that the probe has a permissible range up to 8 μm, which provides a large safety margin for the application of the probe. The difference between two sensitivity curves could be caused by the error of manufacturing and assembly.

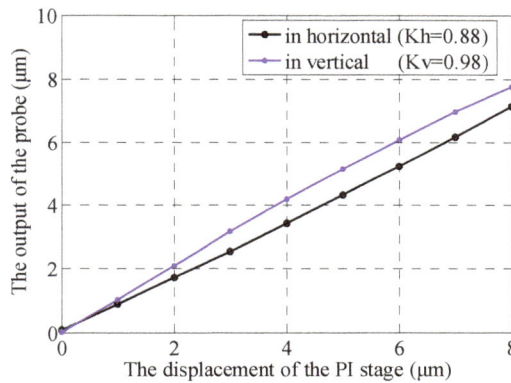

Figure 10. The sensitivity and permissible range of the probe.

4.4. Probe Resolution

An experiment in the same condition as the repeatability test was carried out to investigate the resolution of the probe and the result is shown in Figure 11. The probe ball tip was pushed three steps by the gauge block with a step of 5 nm actuated by the PI stage and then returned back to the initial position. The steps of the probe precisely follow the motion command. Therefore, we can say that the probe has a resolution of less than 5 nm. The different outputs of the probe tip between the beginning and the end might have been caused by the short periodic fluctuation shown in Figure 9.

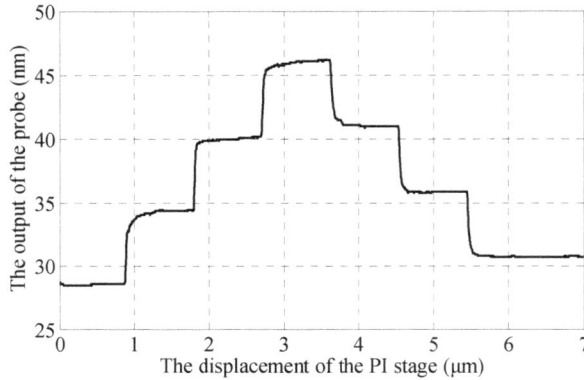

Figure 11. Resolution test results of the probe.

5. Conclusions

A new touch-trigger probe with a small single low-cost sensor for micro-CMMs is presented in this paper. The components and sensing principle of the probe were addressed. The sensitivity of the probe was analyzed, and the uniform sensitivity feature was obtained. The optimal structural parameters conforming to uniform stiffness were obtained. Finite element analysis and experiments were conducted. The probe was verified as having equal stiffness of less than 1 mN/μm in three dimensions, single-direction repeatability of less than 4 nm, resolution better than 5 nm and a small cross-sectional diameter of 40 mm.

Acknowledgments: The reported work is partly funded by the Natural Science Foundation of Anhui Province for Higher Education Institutions (KJ2015A410), the State Key Laboratory of Precision Measuring Technology and Instruments of China (PIL1401), the National Natural Science Foundation of China (51475131, 51475133) and the Foundation of Anhui Electrical Engineering Professional Technique College (2015zdxm05, 2015zdxm06).

Author Contributions: Rui-Jun Li and Kuang-Chao Fan conceived, designed the probe and wrote the paper; Meng Xiang performed the experiments and simulations; Ya-Xiong He manufactured the sensor; Qiang-Xian Huang designed the experiments; Zhen-Ying Cheng analyzed the sensitivity; Bin Zhou created the figures.

Conflicts of Interest: The authors declare no conflict of interest.

Abbreviations

The following abbreviations are used in this manuscript:

CMM   Coordinate measuring machine
QPD   Quadrant photo detector
PI   Physik Instrumente

References

1. McKeown, P. Nanotechnology—Special Article. In Proceedings of the Nano-Metrology in Precision Engineering, Hong Kong, China, 24–25 November 1998; pp. 45–55.
2. Fan, K.C.; Fei, Y.T.; Yu, X.F.; Chen, Y.J.; Wang, W.L.; Chen, F.; Liu, Y.S. Development of a low-cost micro-CMM for 3D micro/nano measurements. *Meas. Sci. Technol.* **2006**, *17*, 524–532.
3. Jäeger, G.; Manske, E.; Hausotte, T. Nanopositioning and Measuring Machine. In Proceedings of the 2nd European Society for Precision Engineering and Nanotechnology, Turin, Italy, 28 May–1 June 2001; pp. 290–293.
4. Takamasu, K.; Furutani, K.R.; Ozono, S. Development of Nano-CMM (Coordinate Measuring Machine with Nanometer Resolution). In Proceedings of the XIV IMEKO World Congress, Tampere, Finland, 1–6 June 1997; pp. 34–39.
5. Peggs, G.N.; Lewis, A.J.; Oldfield, S. Design for a compact high-accuracy CMM. *CIRP Ann.* **1999**, *48*, 417–420.
6. Leach, R.K.; Murphy, J. The Design of Co-Ordinate Measuring Probe for Characterizing Truly Three-Dimensional Micro-Structures. In Proceedings of the 4th EUSPEN International Conference, Glasgow, UK, 30 May–3 June 2004; pp. 230–231.
7. Dai, G.L.; Bütefisch, S.; Pohlenz, F.; Danzebrink, H.U. A high precision micro/nano CMM using piezoresistive tactile probes. *Meas. Sci. Technol.* **2009**, *20*, 1118–1121.
8. Haitjema, H.; Pril, W.O.; Schellekens, P. Development of a silicon-based nano probe system for 3-D measurements. *CIRP Ann.* **2001**, *50*, 365–368.
9. Peiner, E.; Balke, M.; Doering, L.; Brand, U. Tactile probes for dimensional metrology with micro components at nanometre resolution. *Meas. Sci. Technol.* **2008**, *19*, 579–588.
10. Tibrewala, A.; Phataralaoha, A.; Büttgenbach, S. Development, fabrication and characterization of a 3D tactile sensor. *J. Micromech. Microeng.* **2009**, *19*, 125005–125009.
11. Küng, A.; Meli, F.; Thalmann, R. Ultraprecision micro-CMM using a low force 3D touch probe. *Meas. Sci. Technol.* **2005**, *18*, 319–327.
12. Hughes, E.B.; Wilson, A.; Peggs, G.N. Design of high accuracy CMM based on multilateration techniques. *CIRP Ann.* **2000**, *49*, 391–394.
13. Muralikrishnan, B.; Stone, J.A.; Stoup, J.R. Fiber deflection probe for small hole metrology. *Precis. Eng.* **2006**, *30*, 154–164.
14. Kim, S.W. New design of precision CMM based upon volumetric phase-measuring interferometry. *CIRP Ann.* **2001**, *50*, 357–360.
15. Ji, H.; Hsu, H.Y.; Kong, L.X.; Wedding, A.B. Development of a contact probe incorporating a Bragg grating strain sensor for nano coordinate measuring machines. *Meas. Sci. Technol.* **2009**, *20*.
16. Ding, B.Z.; Fei, Y.T.; Fan, K.C. 3D Touch Trigger Probe Based on Fiber Bragg Gratings. In Proceedings of the Metrology, Inspection and Process Control for Microlithography XXIII, San Jose, CA, USA, 19–20 January 2009.
17. Liu, F.F.; Fei, Y.T.; Xia, H.J.; Chen, L.J. A new micro/nano displacement measurement method based on a double-fiber Bragg grating (FBG) sensing structure. *Meas. Sci. Technol.* **2012**, *23*, 54002–54010.

18. Enami, K.; Kuo, C.C.; Nogami, T.; Hiraki, M.; Takamasu, K.; Ozono, S. Development of nano-Probe System Using Optical Sensing. In Proceedings of the IMEKO-XV World Congress, Osaka, Japan, 13–18 June 1999; pp. 189–192.

19. Fan, K.C.; Cheng, F.; Wang, W.L.; Chen, Y.J.; Lin, J.Y. A scanning contact probe for a micro-coordinate measuring machine (CMM). *Meas. Sci. Technol.* **2010**, *21*, 603–616.

20. Chu, C.L.; Chiu, C.Y. Development of a low-cost nanoscale touch trigger probe based on two commercial DVD pick-up heads. *Meas. Sci. Technol.* **2007**, *18*, 1831–1842.

21. Liebrich, T.; Knapp, W. New concept of a 3D-probing system for micro-components. *CIRP Ann.* **2010**, *59*, 513–516.

22. Balzer, F.G.; Hausotte, T.; Dorozhovets, N.; Manske, E.; Jäger, G. Tactile 3D microprobe system with exchangeable styli. *Meas. Sci. Technol.* **2011**, *22*, 94018–94024.

23. Joonh, Y.K.; Jae, W.H.; Yong, S.K.; Lee, D.Y.; Lee, K. Atomic force microscope with improved scan accuracy, scan speed, and optical vision. *Rev. Sci. Instrum.* **2003**, *74*, 4378–4383.

24. Li, R.J.; Fan, K.C.; Huang, Q.X.; Qian, J.Z.; Gong, W.; Wang, Z.W. Design of a large scanning range contact probe for nano-coordinate measurement machines (CMM). *Opt. Eng.* **2012**, *51*, 527–529.

25. Li, R.J.; Fan, K.C.; Miao, J.W.; Huang, Q.X.; Tao, S. An analogue contact probe using a compact 3D optical sensor for micro/nano coordinate measuring machines. *Meas. Sci. Technol.* **2014**, *25*, 1–33.

26. Li, R.J.; Fan, K.C.; Huang, Q.X.; Zhou, H.; Gong, E.M. A long-stroke 3D contact scanning probe for micro/nano coordinate measuring machine. *Precis. Eng.* **2015**, *43*, 220–229.

27. Li, R.J.; Fan, K.C.; Zhou, H.; Wang, N.; Huang, Q.X. Elastic mechanism design of the CMM contact probe. In Proceedings of the SPIE, Chengdu, China, 8–11 August 2013; pp. 182–185.

28. Feng, J.; Li, R.J.; Fan, K.C.; Zhou, H.; Zhang, H. Development of a low-cost and vibration-free constant-temperature chamber for precision measurement. *Sens. Mater.* **2015**, *27*, 329–340.

29. Li, R.J.; Fan, K.C.; Qian, J.Z.; Huang, Q.X.; Gong, W.; Miao, J.W. Stability analysis of contact scanning probe for micro/nano coordinate measuring machine. *Nanotechnol. Precis. Eng.* **2012**, *10*, 125–131.

166

Simulation Model for Correction and Modeling of Probe Head Errors in Five-Axis Coordinate Systems

Adam Gąska, Piotr Gąska and Maciej Gruza

Abstract: Simulative methods are nowadays frequently used in metrology for the simulation of measurement uncertainty and the prediction of errors that may occur during measurements. In coordinate metrology, such methods are primarily used with the typical three-axis Coordinate Measuring Machines (CMMs), and lately, also with mobile measuring systems. However, no similar simulative models have been developed for five-axis systems in spite of their growing popularity in recent years. This paper presents the numerical model of probe head errors for probe heads that are used in five-axis coordinate systems. The model is based on measurements of material standards (standard ring) and the use of the Monte Carlo method combined with select interpolation methods. The developed model may be used in conjunction with one of the known models of CMM kinematic errors to form a virtual model of a five-axis coordinate system. In addition, the developed methodology allows for the correction of identified probe head errors, thus improving measurement accuracy. Subsequent verification tests prove the correct functioning of the presented model.

Reprinted from *Appl. Sci.* Cite as: Gąska, A.; Gąska, P.; Gruza, M. Simulation Model for Correction and Modeling of Probe Head Errors in Five-Axis Coordinate Systems. *Appl. Sci.* **2016**, *6*, 144.

1. Introduction

The development of production engineering and the demand for high quality manufacturing brings about new challenges in the coordinate metrology, which serves as the key tool for quality assurance systems in many branches of industry (automotive, aerospace, machine, *etc.*) [1–3]. With the increase in manufacturing accuracy, the measurements and methods used for assessing measurement uncertainty have to be more precise. The time spent on performing measurements and evaluations also plays a crucial role, as it has a fairly significant impact on the total costs of manufacturing. Therefore, new solutions are being introduced in regards to both the construction of measuring systems, and the methods for the assessment of coordinate measurement accuracy [4,5].

Simulative models constitute one of the newest developments in accuracy estimation. In practice, such models are based on developing virtual models of measuring systems. Their main advantage is the possibility to obtain measurement uncertainty following only a single measurement, whereas the common methods of uncertainty estimation (such as the calibrated workpiece method [6,7], or multiple measurement strategy [8,9]) would require multiple measurements. In some instances, simulative models may also be used for error compensation, and consequently, for improving measurement accuracy.

The virtual Coordinate Measuring Machine (CMM) model was first developed in the 1990s at Physikalisch-Technische Bundesanstalt in Germany [10,11]. The original model was based on the determination of all possible error causes in the entire measuring volume of the CMM, and the evaluation of their impact on measurement results. Thanks to this, the mathematical model of ideal measurement was supplemented by the influence of possible errors that may occur during measurements, and used for performing multiple simulations of the considered measurement. Measurement uncertainty was then determined based on the variability of simulated results and simple statistical operations.

So far, several models have been developed for typical three-axis CMMs. All of those models utilize a similar working principle as the PTB model, and were presented in detail in [12–17]. Lately, a virtual model of measurements performed on the Articulated Arm CMM was also developed [18]. Further works focused on the construction of a virtual model for the Laser Tracker system are currently in progress both in Germany and the UK.

In the last years, it has been possible to observe an increase in the popularity of the so-called five-axis coordinate systems. The five-axis system is created by adding two rotations of a probe head to the standard movements of the machine. Thanks to this solution, it is possible to speed up the measuring process, especially for solids of revolution such as rings, cylinders, spheres, *etc.*, without accuracy loss. Workpieces of this type may be measured using mainly movements of the probe head kinematic pairs. Unlike the three-axis CMM, there is no need to move the whole body of the machine during measurements of all points. The reduction in measurement time is clearly visible, and because of that, the number of installations or retrofits for five-axis systems is becoming a fairly common trend in the industry. Nevertheless, hardly any literature can be found on this subject, and few scientists pay sufficient attention to this type of system. For this reason, the authors decided to investigate the field of errors of probe heads used in five-axis systems, and to attempt to develop a model of this type.

The main sources of errors occurring during coordinate measurements performed on a CMM are kinematic system errors and probe head errors [19–22]. These types of errors are included in all virtual models of CMMs. The models of probe head errors are usually based on error identification with the use of material standards, such as spherical standards and standard rings. Some currently used probe head error simulation models are described below.

A probe head error model utilizing an artificial neural network was developed at the Laboratory of Coordinate Metrology (LCM) of the Cracow University of Technology as part of a Virtual CMM based on assumptions of the matrix method [16]. The model utilizes the Probing Error Function (PEF) described in [13,16]. Probing Error (PE) depends on a number of factors, among them: deformation of the measuring ball during contact, deflection of the stylus under the force of the measurement, probe pretravel, the form error of the measuring ball, non-linearity and differentiation of the transducer characteristics in different directions of deflection, *i.e.*, the directions of probe head movements (u, v, w), reaction to uneven load of the probe head, *etc.* The PE can be defined as a sum of a number of contributing factors. The value of PE changes significantly for different probing directions, so the PEF may be obtained in relation to the probe deflection angle (α). Values of PEF can be determined through measurements of calibrated material standards with known form deviations. The value of the PE at a certain measured point, measured at a specified angle, may be expressed as a radial error obtained during point measurement [16].

The model is divided into two modules: one responsible for systematic error simulation and the second for random error modeling. The ring gauge of a 25 mm diameter has to be measured 32 times, in 64 evenly distributed points, to gather data for model preparation. It was assumed that the 64 measured points are sufficient to model the probe head functioning for any probing direction. On the basis of obtained results, the mean value of the PEF should be attributed to each measured point. These values are treated as a systematic part of the probing system error. The obtained data is used as a learning data set for the constructed artificial neural network, which links the probe head deflection angle and systematic error. The network consists of three layers, and utilizes back propagation and different activation functions (Figure 1). The module was prepared both for inner dimensions—using the gauge ring as a reference—and outer dimensions. In the case of the latter, the entire procedure should be repeated, except that instead of a ring, the spherical standard of a 25 mm diameter should be measured at the reference sphere equator.

The second module is responsible for a simulation of probe head random errors. The chi-squared test should be used to check the distribution of results obtained during the gauge ring measurement. Tests performed previously indicated that the resulting distributions are mostly similar to a normal distribution. In this case, the random error of the probe head connected with a certain probing direction could be modeled using the standard deviation obtained for the considered point. In order to interpolate the values of standard deviation between measured points, the artificial neural network with identical construction to the one in the previous case was used. Finally, a normal distribution generator based on the Monte Carlo method may be used in order to simulate the random error. More details regarding this model may be found in [16].

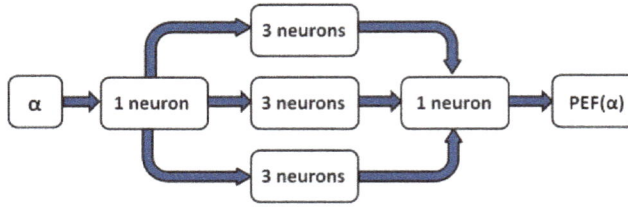

Figure 1. Construction of the neural network used for probe error modeling. α: probe deflection angle; PEF: Probe Error Function.

Another probe head error model developed at LCM is based on the Monte Carlo method. The main idea of the model is the same as in the previously discussed model, namely to link probe head errors with the probing direction; however, this model was adapted to describe the probe spatial approach to the measured surface. In such a situation at least two angles defining tip deflection should be used to sufficiently describe the probe head behavior. Assuming that deflection can be described similarly to the definition of coordinates in a spherical coordinate system, the final probe head error function (PEF) can be written in the following form in Equation (1) [13]:

$$PEF = (\alpha, \beta, PE) \tag{1}$$

The standard used during the model preparation stage is the calibration sphere which fulfills the requirements formulated in [13]. The sphere is measured in evenly distributed points on the upper hemisphere of the standard. These points define the nodes of the reference grid of the model, in which probe head errors are determined empirically through measurements. The arrangement of the points can be described using angles (α, β). The sphere is measured at least 10 times, and on the basis of the acquired results the mean value of the PEF at each point is calculated, as well as standard deviation associated with it. Then, a t-distribution described by $\bar{x}$, s, ν parameters is assigned to each grid node, where $\bar{x}$ is the mean value of the distribution, equal to the mean value of the PEF, s is the dispersion equal to the calculated standard deviation, and ν is the number of degrees of freedom equal to the number of measurements minus one. When the direction is different than the one for which errors were determined experimentally, the bilinear interpolation adapted for the spherical system must be used in order to determine the values describing the probability distribution.

The value of the PE for the chosen probing direction is simulated using the Monte Carlo method. Since the PE can be treated as a vector with the same direction as a probing direction, it is possible to calculate its effect on the coordinates of measured points. The probing errors along axes x, y, z of a machine's datum system (PE_x, PE_y, PE_z) can be expressed as follows:

$$PE_x = PEF \times nx \tag{2}$$

$$PE_y = PEF \times ny \tag{3}$$

$$PE_z = PEF \times nz \tag{4}$$

where:

rx, ny, nz—approach direction cosines
PEF—value of probe head error function

The experiment described in [23] was conducted to check the number of points that should be included in the reference grid. As a result, it was found that the model based on 46 points accurately describes the probe errors. More details regarding this model can be found in [13].

The modules presented above served as a starting point for adapting the probe error function to describe the field of errors of the probe heads utilized on five-axis coordinate systems.

2. Developed Simulative Model and Steps to Its Implementation

This section describes the concept of the developed simulative model, and the experiments that were designed and performed in order to implement this model and verify its reliability.

2.1. Description of Simulative Model

The developed simulative model is based on the concept of PEF described in Section 1. In this case, the number of input quantities that has to be specified in order to obtain the probe error (PE) is equal to three. The PEF may be written as presented in Equation (5):

$$PEF = (A, B, \alpha, PE) \tag{5}$$

where:

A—rotation angle along the horizontal axis of the probe head,
B—rotation angle along the vertical axis of the probe head,
α—angle in which the touch-trigger module is working,
PE—probe error given as a result of PEF usage for considered angles.

The A and B angles were presented in Figure 2. They have their zero positions when the probe is oriented vertically. The α angle is dependent on the A and B angles.

It is defined on the plane perpendicular to the probe when it is oriented using A and B angular positions, and has its zero indication in the direction in which the probe rotates along the A axis in the positive direction. The angle increases in the counterclockwise direction.

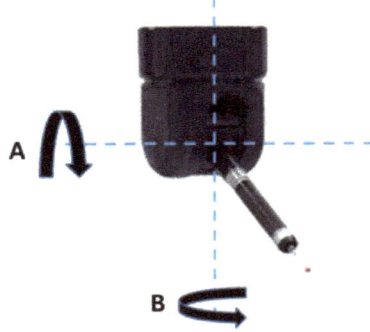

Figure 2. Articulated probe head used in five-axis coordinate systems with revolution axes marked. A: rotation angle along the horizontal axis of the probe head; B: rotation angle along the vertical axis of the probe head.

Depending on those three input parameters, the model provides the output value, which is the simulated value of the PE. The simulation is based on the Monte Carlo method and trilinear interpolation adapted for usage in polar systems. It uses the values of PE identified experimentally for selected A, B and α angles (experiments aimed at determining these values are presented in Section 2.2). The points determined using these angles for the chosen stylus used during measurements may be treated as nodes of the reference point grid. Let A_s, B_s and α_s denote the values of angles for which the simulation of errors has to be performed. A_{s-1}, B_{s-1} and α_{s-1} are the values of angles for the nearest node with angles lower than A_s, B_s and α_s, respectively, and A_{s+1}, B_{s+1} and α_{s+1} are the values of angles for the nearest node with corresponding angles higher than A_s, B_s and α_s. The values of PEs in nodes surrounding the point defined using A_s, B_s and α_s have to be used. In order to simulate the PE value for the point defined using A_s, B_s and α_s, the values of PEs in nodes surrounding this point have to be simulated using the Monte Carlo method. For all of the simulations presented here, the Monte Carlo method uses the scaled and shifted t-distributions with parameters $(\bar{x}, s, \nu)$, where $\bar{x}$ denotes the mean radial PE, s is the standard deviation associated with $\bar{x}$ and ν is the number of degrees of freedom. The parameters of these distributions are determined using the experiment presented in Section 2.2. Hence, the simulation should be performed for nodes $(A_{s-1}, B_{s-1}, \alpha_{s-1})$, $(A_{s-1}, B_{s-1}, \alpha_{s+1})$, $(A_{s-1}, B_{s+1}, \alpha_{s-1})$, $(A_{s-1}, B_{s+1}, \alpha_{s+1})$, $(A_{s+1}, B_{s-1}, \alpha_{s-1})$, $(A_{s+1}, B_{s-1}, \alpha_{s+1})$, $(A_{s+1}, B_{s+1}, \alpha_{s-1})$ and $(A_{s+1}, B_{s+1}, \alpha_{s+1})$. Then, a trilinear interpolation according to Formula (6) should be performed in order to obtain the PE value for the simulated point.

$$PEF(A_s, B_s, \alpha_s) = ((A_{s+1} - A_s)/(A_{s+1} - A_{s-1}) \times ((B_{s+1} - B_s)/(B_{s+1} - B_{s-1}) \times P2 +$$
$$(B_s - B_{s-1})/(B_{s+1} - B_{s-1}) \times P4)) + ((A_s - A_{s-1})/(A_{s+1} - A_{s-1}) \times ((B_{s+1} - B_s)/(B_{s+1} - B_{s-1}) \times P1 + \quad (6)$$
$$(B_s - B_{s-1})/(B_{s+1} - B_{s-1}) \times P3))$$

where:

$P1 = (((\alpha_{s+1} - \alpha_s)/(\alpha_{s+1} - \alpha_{s-1})) \times PEF(A_{s+1}, B_{s-1}, \alpha_{s-1})) + (((\alpha_s - \alpha_{s-1})/(\alpha_{s+1} - \alpha_{s-1})) \times PEF(A_{s+1}, B_{s-1}, \alpha_{s+1}))$

$P2 = (((\alpha_{s+1} - \alpha_s)/(\alpha_{s+1} - \alpha_{s-1})) \times PEF(A_{s-1}, B_{s-1}, \alpha_{s-1})) + (((\alpha_s - \alpha_{s-1})/(\alpha_{s+1} - \alpha_{s-1})) \times PEF(A_{s-1}, B_{s-1}, \alpha_{s+1}))$

$P3 = (((\alpha_{s+1} - \alpha_s)/(\alpha_{s+1} - \alpha_{s-1})) \times PEF(A_{s+1}, B_{s+1}, \alpha_{s-1})) + (((\alpha_s - \alpha_{s-1})/(\alpha_{s+1} - \alpha_{s-1})) \times PEF(A_{s+1}, B_{s+1}, \alpha_{s+1}))$

$P4 = (((\alpha_{s+1} - \alpha_s)/(\alpha_{s+1} - \alpha_{s-1})) \times PEF(A_{s-1}, B_{s+1}, \alpha_{s-1})) + (((\alpha_s - \alpha_{s-1})/(\alpha_{s+1} - \alpha_{s-1})) \times PEF(A_{s-1}, B_{s+1}, \alpha_{s+1}))$

The presented model may serve two primary purposes. The first involves its use as part of virtual model of a five-axis coordinate system. In this application, it should be connected with one of the modules responsible for the simulation of the CMM's kinematic system errors (presented in [10,12,13,16]). For all of the points considered in a simulated measurement, the kinematic errors should first be simulated, and then the probe head errors should be simulated using the presented model. A single simulation of the probe head error for the considered point is done using the methodology presented above. For proper functioning of the virtual model, the simulation should be repeated n times, where n is the number of Monte Carlo trials set by the user or manufacturer of the virtual CMM software.

The second possible usage of the developed model is to utilize it for simulation and correction of probe head errors during the measurements performed in five-axis mode. A detailed description of this process is presented in Section 2.4.

2.2. Implementation Measurements

In order to gather the input data set for model, the reference ring was measured with different angular orientations of the probe head. The reference rings used during experiments were attached to a solid block which was installed in a swivel and tilting vise. A ring with a 20 mm diameter was chosen for the measurements, while other rings were used during datum definition and in further experiments (Section 3.3). In each ring arrangement, the standard was measured 15 times in 64 evenly distributed measured points. For the datum system definition, the CMM works in standard three-axis mode while all reference ring measurements were done using only rotational moves of the probe head. Therefore, the possible influences of machine kinematics are minimized. The positions were selected in a manner that allows covering the majority of the probe head working range. The orientation of the reference ring was changed together with the angular orientation of the probe. The utilized vise allows us to rotate the installed object around two perpendicular axes of revolution, and thus it is possible to set the ring so its axis will be oriented parallel to the stylus in A, B orientation. The authors decided to base the model on measurements carried out in 24 positions defined using B and A probe angles. Angle B changes at $60°$ in the range between $-120°$ and $180°$, while angle A changes at $30°$ in the range between $0°$ and $90°$. Measurements for one of the described positions are presented in Figure 3.

Figure 3. Measurements of standard ring in one of the selected positions.

2.3. Verification Measurements

After the preparation measurements, the ring was measured in further positions to create the verifying data set. The 84 additional positions were selected with B changing at $20°$ between $(-160°; 180°)$, and A changing at $15°$ between $(0°, 90°)$. The same measuring procedure was used. The results obtained during these measurements are used to check the difference between the simulated values and the empirically gathered data. The authors assumed that the PE value may be considered as simulated correctly if it differs from the corresponding value obtained experimentally not more than ± 3 * standard deviation assigned to the considered point.

To perform a simulation and correction of probe head errors during the measurements performed in five-axis mode, the A, B and α angles of all points that were measured have to be recorded. Then the simulations according to the presented methodology have to be performed n times for all of these points. The mean from n simulated values is taken as the probe head error (PE) for the considered point (depending on α). To perform the correction of this error, the PE_x, PE_y and PE_z components have to be calculated using Equations (2)–(4). Then, the values of these components should be subtracted from the actual values of point coordinates x, y, z, giving the corrected point coordinates x_{corr}, y_{corr} and z_{corr} from Equations (7)–(9).

$$x_{corr} = x - PE_x \tag{7}$$

$$y_{corr} = y - PE_y \tag{8}$$

$$z_{corr} = z - PE_z \tag{9}$$

All measured features and relations should be calculated once again using the corrected point coordinates x_{corr}, y_{corr} and z_{corr}. In the presented research, the process of correction of the probe head errors was assisted by a script written in the Python programming language cooperating with macro prepared in Modus software. The raw measurement data including point coordinates, approach vectors and stylus orientation angles was sent to the Python script, which performed the correction of probe head errors and sent back corrected point coordinates to the metrological software, in which the calculation of measured features was done once again.

3 Results

Results presented in this section are the results of experiments described in Section 2. All experiments were performed at the Laboratory of Coordinate Metrology on a Zeiss WMM 850S machine (Carl Zeiss, Jena, Germany) equipped with a Renishaw PH20 probe head and a TP20 standard force module and a stylus with a 4 mm ball and a length of 10 mm. The machine is placed in an air-conditioned room with thermal stability at the level of $20 \pm 0.5\,°C$. Modus metrological software was used on the machine.

3 1. *Results of Identification of Errors*

Figure 4 presents the results for a B angle equal to $120°$ and A changing in the range 0–$90°$.

The results of probe head error identification are used as an input data for the presented simulative model. They are given in tabular form. The example of such a table is presented in Table 1.

175

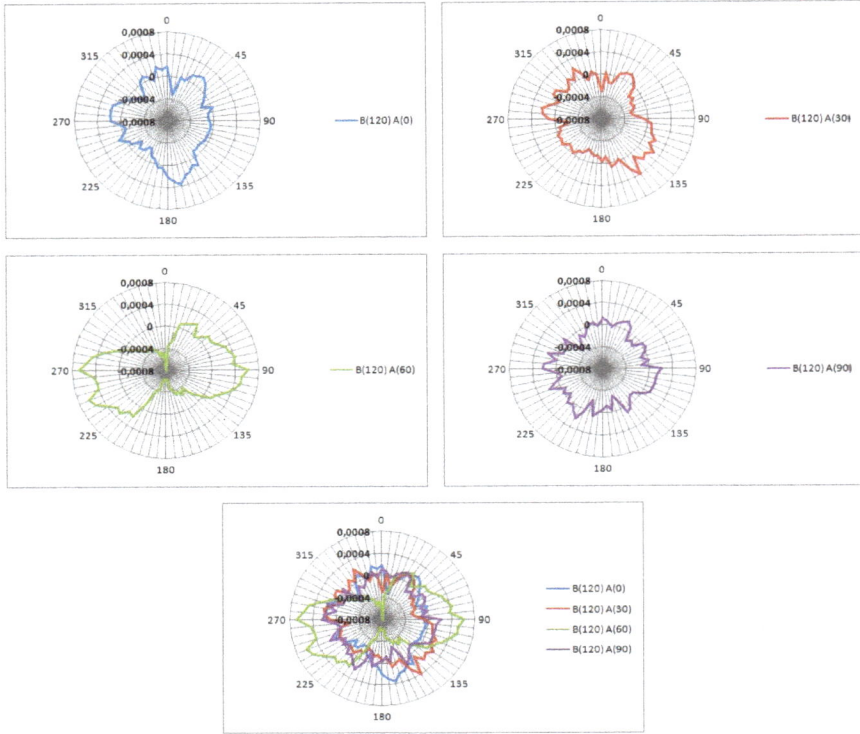

Figure 4. Results of identification of errors for a B angle equal to $120°$ and A changing in the range $0–90°$. All angles given in degrees and errors in mm.

The mean absolute radial error for all positions of the standard ring was equal to 0.0002 mm. The mean value of the standard deviation was equal to 0.0002 mm. The rest of the results of the error identification may be found in the supplementary materials added to this paper (Table S1).

The authors also repeated measurements of the ring in select positions to determine whether the probe error distributions retain their character. These measurements were carried out after the verification measurements to assess how the considerable exploitation of the probe head (the whole measuring procedure for one position consists of 960 measured points) affects the obtained results. Figure 5 presents the results of verification measurement for position $B = 0°$, $A = 45°$ and the results obtained for repeated measurements in the same position. The measurements were repeated after measurements of 46 other positions. The results obtained for the checked positions show satisfactory consistency; therefore, it is justified to base the model on a data set obtained using the experiments described in Section 2.2.

Table 1. Example of input data for simulative model for $A = 30°$ and $B = -120°$. A, B, α given in degrees; $\bar{x}$, s in mm.

A	B	α	x̄	s	A	B	α	x̄	s
30	−120	0.000	−0.00072	0.00048	30	−120	180.000	−0.00017	0.00016
30	−120	5.625	0.00007	0.00019	30	−120	185.625	−0.00034	0.00019
30	−120	11.250	0.00007	0.00020	30	−120	191.250	−0.00024	0.00020
30	−120	16.875	0.00015	0.00019	30	−120	196.875	−0.00034	0.00017
30	−120	22.500	0.00022	0.00013	30	−120	202.500	−0.00010	0.00019
30	−120	28.125	0.00029	0.00016	30	−120	208.125	0.00005	0.00029
30	−120	33.750	0.00015	0.00016	30	−120	213.750	0.00016	0.00027
30	−120	39.375	0.00010	0.00017	30	−120	219.375	0.00032	0.00019
30	−120	45.000	0.00020	0.00023	30	−120	225.000	0.00018	0.00019
30	−120	50.625	0.00019	0.00016	30	−120	230.625	0.00010	0.00012
30	−120	56.250	0.00029	0.00017	30	−120	236.250	0.00012	0.00018
30	−120	61.875	0.00009	0.00021	30	−120	241.875	0.00039	0.00022
30	−120	67.500	0.00010	0.00018	30	−120	247.500	0.00035	0.00020
30	−120	73.125	0.00012	0.00020	30	−120	253.125	0.00026	0.00016
30	−120	78.750	0.00012	0.00016	30	−120	258.750	0.00036	0.00015
30	−120	84.375	0.00016	0.00014	30	−120	264.375	0.00027	0.00019
30	−120	90.000	0.00014	0.00011	30	−120	270.000	0.00009	0.00014
30	−120	95.625	0.00004	0.00010	30	−120	275.625	0.00012	0.00020
30	−120	101.250	−0.00001	0.00023	30	−120	281.250	−0.00003	0.00018
30	−120	106.875	0.00000	0.00012	30	−120	286.875	0.00003	0.00018
30	−120	112.500	−0.00003	0.00015	30	−120	292.500	−0.00001	0.00024
30	−120	118.125	−0.00007	0.00013	30	−120	298.125	−0.00004	0.00016
30	−120	123.750	0.00019	0.00011	30	−120	303.750	−0.00037	0.00024
30	−120	129.375	0.00004	0.00017	30	−120	309.375	−0.00020	0.00018
30	−120	135.000	−0.00004	0.00017	30	−120	315.000	−0.00016	0.00014
30	−120	140.625	−0.00030	0.00019	30	−120	320.625	−0.00006	0.00018
30	−120	146.250	−0.00021	0.00016	30	−120	326.250	−0.00007	0.00013
30	−120	151.875	−0.00023	0.00016	30	−120	331.875	−0.00010	0.00014
30	−120	157.500	−0.00019	0.00017	30	−120	337.500	−0.00032	0.00013
30	−120	163.125	−0.00014	0.00021	30	−120	343.125	−0.00033	0.00015
30	−120	168.750	−0.00009	0.00021	30	−120	348.750	−0.00027	0.00015
30	−120	174.375	−0.00011	0.00011	30	−120	354.375	−0.00021	0.00016

3.2. Results of Model Verification

The verification of the model was done according to methodology presented in Section 2.3. The verification was done for 84 positions of the standard ring. Figures 6 and 7 present the comparison between the results of the experimental probe head error identification and the simulation using the developed model for $A = 75°$ and $B = 80°$ (Figure 5) and $A = -100°$ and $B = 30°$ (Figure 6).

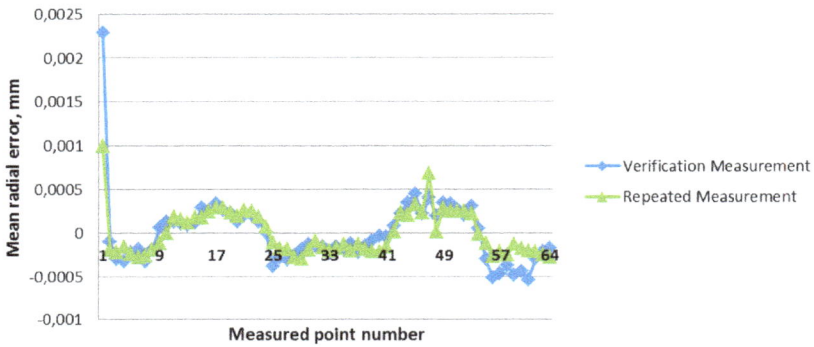

Figure 5. Results of mean Probing Error (PE) values obtained for the chosen position during verification measurements and measurements repeated after a relatively long time of probe head functioning.

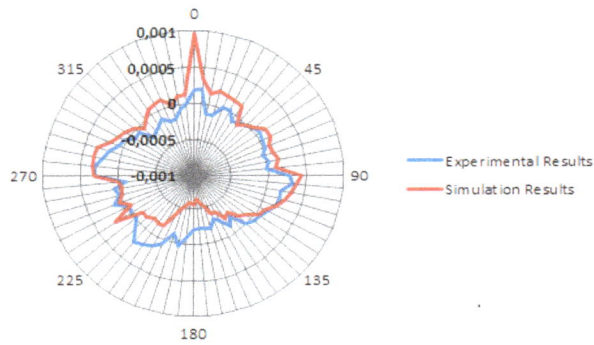

Figure 6. Comparison between results of experimental probe head errors identification and simulation using developed model for $A = 75$ and $B = 80$.

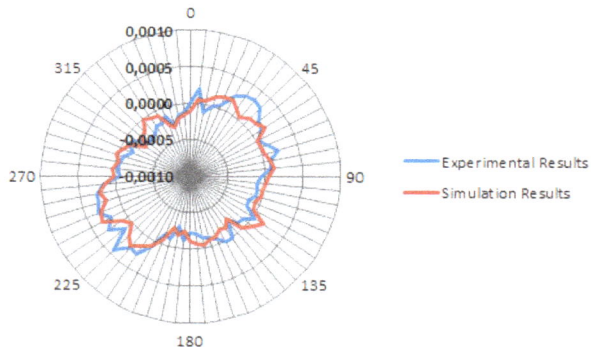

Figure 7. Comparison between results of experimental probe head errors identification and simulation using developed model for $A = -100$ and $B = 30$.

178

For the first position the mean absolute difference between the real values of the probe head errors (determined experimentally) and the simulated ones was equal to 0.0002 mm and for the second position 0.0001 mm. The authors assumed that the PE value may be considered as simulated correctly if it differs from the corresponding value obtained experimentally not more than $\pm 3 \times$ standard deviation assigned to the considered point. Out of 64 simulated values of PE for each position, 61 (95.31%) were properly simulated for the first position and 64 (100%) for the second position (according to assumption presented in Section 2.3).

3.3. Example of Probe Head Error Correction

The possibility of correcting probe head errors was presented on the example of measurements of the standard ring with a diameter of 34 mm. The values of diameter and roundness deviation were determined during the measurements. The ring was mounted in the plane perpendicular to the probe oriented using $A = 43°$ and $B = 91°$. Table 2 presents the results obtained during the real measurements without compensation of the probe head errors and after the compensation done using the methodology presented in Section 2.4. Results were compared with the data given on the calibration certificate of the ring.

Table 2. Results of roundness deviation measurements (given in mm) for standard ring with and without probe head errors correction.

Number of Points Used	Without Correction	With Correction	Calibration Certificate
64	0.0013	0.0010	0.0004
16	0.0012	0.0008	0.0004
8	0.0010	0.0005	0.0004

4. Discussion

The primary conclusion of the presented experiments is that the developed simulation model allows for faithful reproduction of the error field for probe heads used in five-axis coordinate systems. In the case of the two presented positions used during verification measurements, the percentage of properly simulated error values was 95.31% and 100%, respectively. Such high numbers prove the effectiveness of the presented model and suggest its possible application as a module for probe head error simulation in virtual models of five-axis CMMs.

The developed model may also be helpful during the correction of probe head errors. As was shown in Table 2, the precision of the roundness deviation measurements has noticeably improved following the implementation of the correction using the methodology presented in Section 2.4. It should be noted that the effectiveness of probe head error correction is more pronounced in the case of a smaller number of measuring points used for the calculation of geometrical features. In this paper, example measurements were carried out on a circle, but the authors' previous research suggests similar relationships in the case of other features. This phenomenon is connected with the averaging tendency of the least squares method, usually utilized for calculating features.

The results presented in Figure 5 indicate good stability of probe head error characteristics. It can be concluded from this that irrespective of the usage of the developed model, the initial measurements aimed at gathering input data for the model would not have to be repeated very frequently. The ideal interval for the re-identification of probe head errors should be investigated for different probe head types, but the authors suppose that it could be expressed in months.

As for future research directions, the most important goal is to develop a fully functional virtual model of a five-axis coordinate system. The research presented in this paper marks the first step in this direction. The simulative model discussed here should be connected to a model responsible for simulating kinematic system errors of the machine. Such a combined virtual model would reduce the time spent on the determination of the uncertainty of measurements performed on the five-axis coordinate system. Seeing as numerous systems of this type are presently used in industrial conditions, this would be likely to significantly reduce costs associated with quality control, and as a natural consequence, the total production costs would also drop down.

The model presented in this paper was checked for the measurements of regular rotary features for which the probes such as the one discussed here are usually applied in practice. In case of features such as cylinders, circles, spheres, cones, *etc.*, the measurements may be easily performed using only the movements of the probe head, which significantly reduces their total time (in relation to the standard three-axis CMM). The measurements of more complicated shapes are also possible but the reduction of the measurement duration is not that meaningful as in the case of regular features. So it may be concluded that the model works properly for the majority of practical usages of the discussed probes. However, in the future, it will be advisable to also include in it the possibility of simulating the measurements of more complicated geometries, when the stylus has to move on complex trajectories and the α angle cannot be easily defined on the plane perpendicular to the probe when it is oriented using A and B angular positions.

The next important possibility of improving the presented model is to reduce the number of orientations and points used during identification measurements. It should be possible to minimize their number in a way that simultaneously reduces the time needed for model implementation, and provides faithful reproduction of a real field of probe head errors.

Supplementary Materials: The following are available online at www.mdpi.com/2076-3417/6/5/144/s1, Table S1: Complete results of probe head errors identification.

Acknowledgments: Reported research was realized as part of a project financed by National Science Centre, Poland, grant No. 2015/17/D/ST8/01280. Costs to publish in open access were also paid from this source of funding.

Author Contributions: Adam Gąska and Piotr Gąska conceived and designed the simulative model and the experiments; Piotr Gąska and Maciej Gruza performed the experiments; Piotr Gąska prepared Python scripts and Modus macros; Piotr Gąska and Adam Gąska analyzed the data; Adam Gąska and Piotr Gąska wrote the paper.

Conflicts of Interest: The authors declare no conflict of interest.

Abbreviations

The following abbreviations are used in this manuscript:

CMM	Coordinate Measuring Machine
LCM	Laboratory of Coordinate Metrology
PEF	Probe Error Function
PE	Probe Error

References

1. Kunzmann, H.; Pfeifer, T.; Schmitt, R.; Schwenke, H.; Weckenmann, A. Productive metrology-adding value to manufacture. *CIRP Ann. Manuf. Technol.* **2005**, *54*, 691–704.
2. Caja, J.; Gomez, E.; Maresca, P. Optical measuring equipments. Part I: Calibration model and uncertainty estimation. *Precis. Eng.* **2015**, *40*, 298–304.
3. Asano, Y.; Furutani, R.; Ozaki, M. Verification of interim check method of CMM. *Int. J. Autom. Technol.* **2011**, *5*, 115–119.
4. Cuesta, E.; Alvarez, B.; Sanchez-Lasheras, F.; Gonzalez-Madruga, D. A statistical approach to prediction of the CMM drift behaviour using a calibrated mechanical artefact. *Metrol. Meas. Syst.* **2015**, *22*, 417–428.
5. Takamasu, K.; Takahashi, S.; Abbe, M.; Furutani, R. Uncertainty estimation for coordinate metrology with effects of calibration and form deviation in strategy of measurement. *Meas. Sci. Technol.* **2008**, *19*.
6. International Organization for Standardization. *ISO 15530-3:2011—Geometrical Product Specifications (GPS)—Coordinate Measuring Machines (CMM): Technique for Determining the Uncertainty of Measurement—Part 3: Use of Calibrated Workpieces or Measurement Standards*; ISO: Geneva, Switzerland, 2011.
7. Weckenmann, A.; Lorz, J. Monitoring coordinate measuring machines by calibrated parts. *J. Phys. Conf. Ser.* **2005**, *13*, 190–193.
8. Sato, O.; Osawa, S.; Kondo, Y.; Komori, M.; Takatsuji, T. Calibration and uncertainty evaluation of single pitch deviation by multiple-measurement technique. *Precis. Eng.* **2010**, *34*, 156–163.

9. Osawa, S.; Busch, K.; Franke, M.; Schwenke, H. Multiple orientation technique for the calibration of cylindrical workpieces on CMMs. *Precis. Eng.* **2005**, *29*, 56–64.

10. Trapet, E.; Franke, M.; Hartig, F.; Schwenke, H.; Wadele, F.; Cox, M.; Forbers, A.; Delbressine, F.; Schnellkens, P.; Trenk, M.; *et al. Traceability of Coordinate Measuring Machines According to the Method of the Virtual Measuring Machine*; PTB: Braunschweig, Germany, 1999.

11. Schwenke, H.; Siebert, B.R.L.; Wäldele, F.; Kunzmann, H. Assessment of uncertainties in dimensional metrology by Monte Carlo simulation: Proposal of a modular and visual software. *CIRP Ann. Manuf. Technol.* **2000**, *49*, 395–398.

12. Wilhelm, R.G.; Hocken, R.; Schwenke, H. Task specific uncertainty in coordinate measurement. *CIRP Ann. Manuf. Technol.* **2001**, *50*, 553–563.

13. Sładek, J.; Gąska, A. Evaluation of coordinate measurement uncertainty with use of virtual machine model based on Monte Carlo method. *Measurement* **2012**, *45*, 1564–1575.

14. Aggogeri, F.; Barbato, G.; Barini, E.M.; Genta, G.; Levi, R. Measurement uncertainty assessment of coordinate measuring machines by simulation and planned experimentation. *CIRP J. Manuf. Sci. Technol.* **2011**, *4*, 51–56.

15. Giusca, C.L.; Leach, R.K.; Forbes, A.B. A virtual machine-based uncertainty evaluation for a traceable areal surface texture measuring instrument. *Measurement* **2011**, *44*, 988–993.

16. Sładek, J. *Coordinate Metrology: Accuracy of Systems and Measurements*; Springer: Berlin, Germany, 2016.

17. Maresca, P.; Gomez, E.; Caja, J.; Barajas, C. Virtual environment for the simulation of a horizontal coordinate measuring machine in the teaching of dimensional metrology. *Procedia Eng.* **2013**, *63*, 234–242.

18. Ostrowska, K.; Gąska, A.; Sładek, J. Determining the uncertainty of measurement with the use of a Virtual Coordinate Measuring Arm. *Int. J. Adv. Manuf. Technol.* **2014**, *71*, 529–537.

19. Cuesta, E.; Talenti, A.; Patino, H.; Gonzalez-Madruga, D.; Martínez-Pellitero, S. Sensor prototype to evaluate the contact force in measuring with coordinate measuring arms. *Sensors* **2015**, *15*, 13242–13257.

20. Nafi, A.; Mayer, J.R.R.; Wozniak, A. Novel CMM-based implementation of the multi-step method for the separation of machine and probe errors. *Precis. Eng.* **2011**, *35*, 318–328.

21. Gąska, A.; Krawczyk, M.; Kupiec, R.; Ostrowska, K.; Gąska, P.; Sładek, J. Modeling of the residual kinematic errors of coordinate measuring machines using Laser Tracer system. *Int. J. Adv. Manuf. Technol.* **2014**, *73*, 497–507.

22. Dobosz, M.; Woźniak, A. CMM touch trigger probes testing using a reference axis. *Precis. Eng.* **2005**, *29*, 281–289.

23. Gąska, A.; Harmatys, W.; Gąska, P.; Gruza, M.; Mathia, T.; Sładek, J. Optimization of probe head errors model used in Virtual CMM systems. In Proceedings of the 11th International Conference on Laser Metrology and Machine Performance (LAMDAMAP) 2015, Huddersfield, UK, 17–18 March 2015; Blunt, L., Hansen, H.N., Eds.; euspen: Cranfield, UK, 2015.

A Model to Determinate the Influence of Probability Density Functions (PDFs) of Input Quantities in Measurements

Jesús Caja, Piera Maresca and Emilio Gómez

Abstract: A method for analysing the effect of different hypotheses about the type of the input quantities distributions of a measurement model is presented here so that the developed algorithms can be simplified. As an example, a model of indirect measurements with optical coordinate measurement machine was employed to evaluate these different hypotheses. As a result of the different experiments, the assumption that the different variables of the model can be modelled as normal distributions is proved.

Reprinted from *Appl. Sci.* Cite as: Caja, J.; Maresca, P.; Gómez, E. A Model to Determinate the Influence of Probability Density Functions (PDFs) of Input Quantities in Measurements. *Appl. Sci.* **2016**, *6*, 190.

1. Introduction

The determination measurement's uncertainty made with coordinates measuring machines (CMMs) is an important line of research in the field of coordinate metrology [1,2].

Although there are different papers [3,4] in which the uncertainty calculation is based on GUM approach [5], this model is impractical for most situations that appear in the field of coordinate metrology, since the GUM approach does not provide a solution to the singularities that arise, for example,

- multi-dimensional models based on the coordinates of multiple points;
- the impossibility of determining the sensitivity coefficients of some parameters when these are the result of the application of filters, adjustment algorithms, etc.;
- the existence of input variables of the model that have non-symmetric distributions;
- the nonlinearity of calculation models that force one to consider higher-order terms. Wilhelm et al. [6] conducted a comprehensive analysis of the various models for calculating.

To avoid these problems, different authors have proposed different methods. Wilhelm et al. [6] used a numerical method for calculating uncertainties associated with specific measurement tasks, due to the complexity of the measures. They analysed the concept of *"virtual CMM"*, which materialises a very precise mathematical model of behavior of a CMM. This model simulates data acquisition, measurement strategy, and physical behavior of the CMM. By the later use of the Monte Carlo method, it is possible to determine the measurement uncertainty. Abbe et al. [7] proposed a method for the evaluation of the uncertainty associated with measurements made by a CMM from information obtained after checking the machine using the ISO 10360-2 [8].

The Monte Carlo method [9] is the appropriate solution to estimate the uncertainty associated with measurements obtained with equipment based in coordinate metrology and allows solving problems such as:

- arbitrarily complicated models or input quantities of the model present asymmetrical probability density functions (PDF).

According to Supplements 1 and 2 to the GUM [9,10], the steps to be followed to use the Monte Carlo method are as follows:

- formulation, in which the following is defined: (1) the output quantities of model (measurand), (2) the input quantities of the model, (3) the measurement model that connects the inputs with the outputs, (4) the assignment of PDF to each input quantity;
- the propagation of the PDF assigned to the input quantities using the measurement model, obtaining the PDF of each output quantity; and
- a summary determining for each output quantity, from its PDF, their mathematical expectation, standard uncertainty, uncertainty coverage interval, and covariance matrix.

It is possible to find metrological models that provide correlated output quantities with PDF not comparable to any known. When these quantities are used in subsequent calculation measurement models, the propagation of these quantities cannot use the procedures described in Supplements 1 and 2 of the GUM. In these cases, the theory of copulas is often used [11–14], which develops functions capable of describing dependencies between variables and providing multivariate data that model correlated distributions. The use of the theory of copulas has disadvantages such as high computation times and complexity of the developed algorithms, disadvantages that in an industrial environment are critical parameters.

The aim of this paper is to analyse the effect of various hypotheses about the type of the input quantities distributions of the measurement model so that the developed algorithms can be simplified. A model of indirect measurements with optical CMMs is employed, and three hypotheses effects on the measurement results (distance between graduations on a line scale and its associated uncertainty) are studied:

- the use of the theoretical distributions, obtained in the determination of the calibration parameters of the equipment;

- the simplification of the known theoretical distributions (*t*-student, uniform, Weibull, etc.) that better model the behaviour of the theoretical distributions are used; and
- the simplification of the theoretical distributions with normal distributions.

For this measurement model, the input quantities studied are the parameters that assure the metrological traceability of the equipment.

2. Materials and Methods

In this section, a mathematical model is presented to characterize the vision subsystem of the CMM. With a developed calibration procedure [15], it is possible to calculate the uncertainties of the measurements made therewith [16].

2.1. Optical Coordinate Measurement Machine Model

The vision subsystem model of the optical coordinate measurement machine used in this paper is the so-called affine camera, in which the optical centre is located at an infinity point [17]. This model is used to model systems with telecentric lenses/lens systems [18], allowing for the transformation of the coordinates of a point in space (3D) called *"coordinates of the world system"* into the coordinates of a point of an image (2D) called *"coordinates of the image system digitized"*. This inverse transformation is represented in the matrix form by:

$$\left. \begin{array}{l} \mathbf{X}'_{hm} = \mathbf{Q}' \cdot \mathbf{U}_{hpi} \\ \mathbf{X}'_{hm} = [\mathbf{R}' \mid \mathbf{t}]^{-1} \cdot \mathbf{K}^{-1} \cdot \mathbf{U}_{hpi} \end{array} \right\} \quad \text{with} \quad \mathbf{K}^{-1} = \begin{pmatrix} 1/\alpha_x & -s/\left(\alpha_x \cdot \alpha_y\right) & -\left(\alpha_y \cdot x_0 + s \cdot y_0\right)/\left(\alpha_x \cdot \alpha_y\right) \\ 0 & 1/\alpha_y & -y_0/\alpha_y \\ 0 & 0 & 1 \end{pmatrix}, \quad (1)$$

where $\mathbf{X}'_{hm} = (x_m, y_m, 1)^T$ are the coordinates of a point in the world system, and] are the pixel coordinates of the digitized image system. $\mathbf{R}'$ is an orthogonal rotation matrix 3×3, and $\mathbf{t}$ is a translation matrix 3×1. With these two matrixes, the position of the camera with respect to the world system is defined. s characterizes the possibility of the lack of perpendicularity between the axes, x_0, y_0 represent the image centre coordinates, α_x, α_y model the camera pixels size along the X- and Y-axes, and $\delta_{u'}$, $\delta_{v'}$ represent the geometric distortion of the coordinates u' and v':

$$\delta_{u'} = \overbrace{k_1 \Delta u' \left(\Delta u'^2 + \Delta v'^2\right) + k_2 \Delta u' \left(\Delta u'^2 + \Delta v'^2\right)^2}^{Radial} + \overbrace{p_1 \left(3\Delta u'^2 + \Delta v'^2\right) + 2p_2 \Delta u' \Delta v'}^{Decentering} + \overbrace{s_1 \left(\Delta u'^2 + \Delta v'^2\right)}^{Thin\ prism}, \quad (2)$$
$$\delta_{v'} = k_1 \Delta v'(\Delta u'^2 + \Delta v^2) + k_2 \Delta v'\left(\Delta u'^2 + \Delta v'^2\right)^2 + 2p_1 \Delta u' \Delta v' + p_2(\Delta u'^2 + 3\Delta v'^2) + s_2(\Delta u'^2 + \Delta v'^2)$$

where $\Delta u' = u' - x_0$, $\Delta v' = v' - y_0$ and $k_1, k_2, p_1, p_2, s_1, s_2$ represent the geometric distortion coefficients.

By employing a fixed frequency grid distortion target, it is possible to calibrate the instrument vision subsystem [15] so that the traceability of the measures subsequently made by the equipment is ensured. The results of the calibration of a MITUTOYO equipment, model Ultra-QV350 (measuring range ($X \times Y \times Z$): 350 mm × 350 mm × 150 mm, resolution: 0.01 µm) available in the Centro Español de Metrología (CEM) (Tres Cantos, Spain) are shown below. The distortion target images (dot diameter: 65 µm and step: 125 µm) have been obtained by using diascopic illumination at 20% of its nominal value, with a magnification of 10×, which determines a nominal pixel size of 0.89 µm. The temperature during the measurement time is kept in the range 20 ± 0.2 °C. Table 1 shows the results of the simulations performed for a number of replications $M = 10^4$ (the Monte Carlo method is used for determining the uncertainty associated with the model calibration parameters).

Table 1. Results of the calibration parameters of the optical coordinate measuring machine.

Parameter	Parameter Estimation y_i	Standard Uncertainty $u(y_i)$	Shortest 95% Coverage Interval	
			Lower Limit	Upper Limit
$\mathbf{K}^{-1}(1,1)$ (mm/pixel)	9.855×10^{-4}	2.5×10^{-6}	9.803×10^{-4}	9.912×10^{-4}
$\mathbf{K}^{-1}(1,2)$ (mm/pixel)	9.662×10^{-7}	4.8×10^{-9}	9.560×10^{-7}	9.768×10^{-7}
$\mathbf{K}^{-1}(1,3)$ (mm)	-0.31610	0.00099	-0.31792	-0.31434
$\mathbf{K}^{-1}(2,2)$ (mm/pixel)	-9.803×10^{-4}	2.5×10^{-6}	-9.856×10^{-4}	-9.751×10^{-4}
$\mathbf{K}^{-1}(2,3)$ (mm)	0.23576	0.00076	0.23446	0.23705
u_r (pixel)	320.50	0.61	320.35	320.65
v_r (pixel)	240.49	0.51	240.34	240.64
k_1 (pixel^{-2})	0.1×10^{-8}	8.1×10^{-8}	-1.77×10^{-7}	1.76×10^{-7}
k_2 (pixel^{-4})	0.4×10^{-13}	6.2×10^{-13}	-1.27×10^{-12}	1.33×10^{-12}
s_1 (pixel^{-1})	-0.1×10^{-6}	2.3×10^{-6}	-4.5×10^{-6}	4.4×10^{-6}
s_2 (pixel^{-1})	0.3×10^{-6}	2.3×10^{-6}	-4.1×10^{-6}	4.7×10^{-6}
p_1 (pixel^{-1})	-0.4×10^{-6}	6.8×10^{-6}	-1.34×10^{-5}	1.32×10^{-5}
p_2 (pixel^{-1})	-0.1×10^{-6}	5.3×10^{-6}	-1.1×10^{-6}	1.0×10^{-6}

Since the above variables have common input variables, there are correlated [15]. Figure 1a,b show, as an example, some of the histograms of the output variables of the calibration model. From his analysis, and from the histograms of the variables not shown, it is possible to find that these variables could be assimilated reasonably well to the following distributions:

- t-student, with different degrees of freedom (DOF), variables, $\mathbf{K}^{-1}(1,1)$, $\mathbf{K}^{-1}(1,2)$, $\mathbf{K}^{-1}(1,3)$, $\mathbf{K}^{-1}(2,3)$, $\mathbf{K}^{-1}(2,3)$, k_1, and k_2;
- normal, variables p_1, p_2, s_1, and s_2; and
- other types, variables u_r and v_r. If these distributions are analysed, it is found that 99% of their values could be assimilated reasonably well to a rectangular distribution.

The u_r and v_r variables have a high concentration of values in a relatively narrow range and very extensive distribution tails. This behaviour is due to the fact that the geometric distortion parameters obtained are practically zero, independently of the value taken by the geometric distortion centre; within limits, the corrections obtained with these parameters are virtually null.

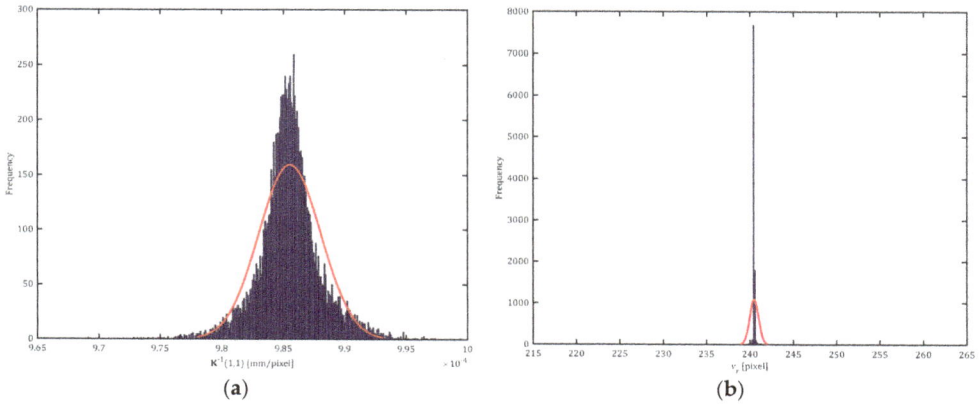

Figure 1. (a) $K^{-1}(1, 1)$ histogram; (b) v_r histogram.

2.2. Considerations about the Types of Distributions of the Parameters of the Vision Subsystem

As noted above, the vision subsystem parameters can be assimilated to some of the known types of distributions in a reasonably correct way. In this subsection, the effect that occurred as a result of a measurement (values of the parameter and its uncertainty) when different assumptions about the type of distribution that match the parameters of vision subsystem are made are analysed. To do this, for example, a measurement model that assesses the distance between graduations of a glass line scale and its associated uncertainty using the Monte Carlo method is implemented. The three hypotheses indicated in the introduction are considered.

For Hypotheses 2 and 3, for each vision subsystem parameter, and from the information obtained from its histogram, the mean value, standard deviation, and number of degrees of freedom (*t*-student distribution) are determined if necessary.

2.2.1. Procedure for Measuring a Glass Line Scale

A NPL (National Physical Laboratory) glass line scale (Figure 2a), with a nominal length of 100 nm and a distance of 0.1 mm between graduations, was used. Its main purpose was to serve as a high accuracy standard for performing calibrations in the industrial area. The graduation width was about 10 μm. The scale contained two parallel horizontal lines, about 50 μm apart (Figure 2a), at the start and

187

the end of the scale, used as an alignment axis. The coefficient of linear expansion of the scale, as indicated by the manufacturer, is: $\alpha = (0.5 \pm 0.1) \times 10^{-6}\,\text{K}^{-1}$.

The glass line scale was calibrated by the CEM using a microscope-CCD camera assembly and laser interferometric system. The results up to a nominal position of 0.4 mm were as follows (Table 2).

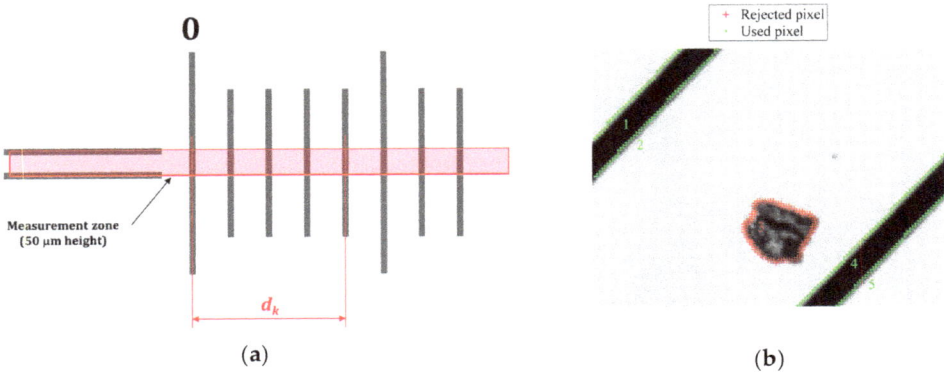

(a)

(b)

Figure 2. (a) Glass line scale. (b) Edges detected by processing the image of the line scale.

Table 2. Calibration data of the glass line scale.

Nominal Position (mm)	Measured Valued (mm)	Deviation from Nominal Position (µm)	Expanded Uncertainty ($k = 2$) (µm)
0	0.00000	0.00	0.08
0.1	0.10003	0.03	0.08
0.2	0.20006	0.06	0.08
0.3	0.30001	0.01	0.08
0.4	0.39999	−0.01	0.08

The measurands are the distances d_k between the reference centreline of the zero graduation and the centreline of the k-th graduation. The measurements were made on the section of the scale to fall between the two horizontal lines used as an alignment axis. The field of vision of the Ultra-QV350 equipment for the measurement conditions used above (magnification $10\times$) was 0.62 mm $\times$ 0.47 mm. Therefore, the distances between graduations of the glass line scale are measured up to the nominal position 0.4 mm so that it is possible to compare the results obtained in different positions.

This phase was divided into the following steps:

- acquisition of multiple images of the glass line scale. Ten images at intervals of 30 s were captured, with the standard in the same position, and the information obtained from the images was used in the subsequent calculation of uncertainties;
- correction and normalization of the luminance of the image [19];
- detection of the pixels that formed the edge of the glass line scale graduations (Figure 2b), employing the Canny edge detection method with "global thresholding". Each k-th graduation of the glass line scale, with $k = 0, \ldots, n_t$, has two edges, called an odd edge and even, the edges of a k-th graduation are characterized by a several number of points $(p_{k_{odd}}, p_{k_{even}})$, and, for each of the k graduations, the pixels of the odd edge $\left(u_i^k, v_i^k \right) \Big|_{i=1}^{p_{k_{odd}}}$ and the even edge $\left(u_j^k, v_j^k \right) \Big|_{j=1}^{p_{k_{even}}}$ are obtained;
- conversion of the previous coordinates (in pixels) of the edge in units of length. The above relationship is determined by the expression:

$$\mathbf{X}_m = \mathbf{K}^{-1} \cdot \mathbf{U}_m, \tag{3}$$

where $\mathbf{U}_m = \left(u_m' - \delta_u', v_m' - \delta_v' \right)$ and $\mathbf{X}_m = (x_m, y_m, 1)^T$ are the coordinates in units of length. The components of the matrix $\mathbf{K}^{-1}$, the terms of geometric distortion (distortion coefficients $k_1, k_2, p_1, p_2, s_1, s_2$, and the geometric distortion centre u_r, v_r), were obtained in the previous calibration of the vision subsystem (Section 2.1).

- determination of the distance between graduations. Once the coordinates of the edge pixels have been processed, the distances d_k between the zero graduation and the k-th [16] are determined).

2.2.2. Model for Calculating Uncertainties

As can be deduced from the scientific literature, the Monte Carlo method is fit to determine the uncertainty of the previous values d_k distances. This method is divided into the following steps that were particularized for our mathematical model:

1. Definition of output variables: the distance d_k between the zero gradation and the line k-th graduation.
2. Definition of input quantities: determining the d_k distances has the following input variables: the light intensity of the pixels $I(i,j)$ of the image of the glass line scale and the results of the calibration of the vision subsystem materialized by $\mathbf{K}^{-1}, u_r, v_r, k_1, k_2, s_1, s_2, p_1, p_2$, i.e., the inverse calibration camera matrix, the centre of geometric distortion and geometric distortion coefficients. If the results are corrected for the effect of temperature, it is necessary to employ the linear

expansion coefficient α of the glass line scale and the temperature difference of the glass line scale ΔT with respect to the reference.

3. Assignment of the PDF to the input variables: The uncertainty associated with the light intensity of the pixels is obtained according to the work of De Santo et al. [20]. The uncertainty of the variables is determined by the hypothesis considered:

- Hypotheses 1 and 2: It can be difficult to generate random data with dependence when they have distributions that are not from a standard multivariate distribution. Indeed, some of the standard multivariate distributions can model only some limited types of dependence. In these cases, copulas are often used [21]. These functions can describe dependencies between variables and provide distributions that model correlated multivariate data. Its use allows the construction of bivariate or multivariate distribution by specifying marginal univariate distributions. For this, it is necessary to choose the type of copula so as to allow generating a correlation structure between variables. To simulate multivariate variables in this paper the following steps are followed:

  ○ Simulation of a multivariate normal with zero mean and covariance matrix unit.

  ○ Calculation of a multivariate normal with zero mean and covariance matrix $\mathbf{U_C}$, which is the covariance matrix of the calibration results of the vision subsystem, Equation (4):

$$\mathbf{U_C} = \begin{pmatrix} u^2\left(\mathbf{K}^{-1}(1,1)\right) & u^2\left(\mathbf{K}^{-1}(1,1),\mathbf{K}^{-1}(1,2)\right) & \cdots & u^2\left(\mathbf{K}^{-1}(1,1),p_2\right) \\ u^2\left(\mathbf{K}^{-1}(2,1),\mathbf{K}^{-1}(1,1)\right) & u^2\left(\mathbf{K}^{-1}(1,2)\right) & \cdots & u^2\left(\mathbf{K}^{-1}(1,2),p_2\right) \\ \vdots & \vdots & \ddots & \vdots \\ u^2\left(p_2,\mathbf{K}^{-1}(1,1)\right) & u^2\left(p_2,\mathbf{K}^{-1}(1,2)\right) & \cdots & u^2(p_2) \end{pmatrix}. \quad (4)$$

  ○ Prior transformation of normal uniforms, whose values are contained in the interval [0, 1].

  ○ It is possible to obtain the variable c_i from the variable using the inverse function $F^{-1}(u_i)$:

$$c_i = F^{-1}(u_i). \quad (5)$$

  In the case of Hypothesis 1, the probability distribution function $F(x)$ is estimated using a kernel density estimation (KDE) [22,23]; in the case of Hypothesis 2, the probability distribution function $F(x)$ is a known, univariate marginal distribution (normal, t-student, ...).

- Hypothesis 3: The uncertainty of the variables $\mathbf{c} = \left(\mathbf{K}^{-1}, u_r, v_r, k_1, k_2, s_1, s_2, p_1, p_2\right)$ is calculated from the covariance matrix obtained in the calibration of the vision subsystem. As noted in Section 2.1, these variables are correlated. This hypothesis assumes that the above variables have PDF equivalent to Gaussian distributions. Taking into account the recommendations of Supplement 1 to the GUM, the determination of the PDF of these variables uses the method of multivariate normal distribution [9]. Multivariate normal distribution $\mathbf{N}\left(\mathbf{c}, \mathbf{U_C}\right)$ is assigned to variables $\mathbf{c}$.

4. Propagation: considering that the calculation model is relatively complex, the recommendation of Supplement 1 to the GUM, Section 7.2.3 [9], is taking into account. In response to this recommendation, the model is replicated a number of times equal to 10^4.

5. Summary: the statistics variable mean and standard deviation of the values obtained in the simulations shall be calculated. To calculate the coverage interval, the shortest interval method is used [24].

3. Results

Once the model for calculating the distances d_k and for estimating its associated uncertainties was defined, the measurement of the glass line scale was performed in a position approximately parallel to the X axis of the machine.

The measurement was performed with the optical equipment Ultra-QV350, whose characteristics have already been defined in Section 2.1. The glass line scale images were obtained by using transmitted illumination at 20% of its nominal value with $10\times$ magnification. The glass line scale temperature during measurement remained in the range $20\,^\circ\mathrm{C} \pm 0.1\,^\circ\mathrm{C}$. Thus, no temperature correction is applied to the distances d_k.

Table 3 shows the results obtained by measuring the glass line scale considering the Hypothesis 1. Figure 3a,b show the histograms of the distances d_2 and d_4. These results obtained by developing the Hypotheses 2 and 3 are shown in Table 4, Figure 4a,b and Table 5, Figure 5a,b, respectively. Analyzing the results of the above tables, it is observed that variations within a single estimate of the measurand d_k obtained in the various hypotheses are between 1 nm and 22 nm, values that can be considered negligible considering that the nominal size pixel in the measurement conditions used was 0.98 μm.

By performing the same analysis for the standard uncertainty of the distances d_k, it was found that it varied between 26 nm and 74 nm. These variations were greater as distance d_k increased. When the results of Hypotheses 1 and 3 were compared, it was determined that its variation was between 1 nm and 30 nm. As in the previous case, these variations can be considered negligible.

Table 3. Measurement results of the glass line scale in horizontal position. Hypothesis 1.

Distances	Estimate d_k (mm)	Deviations to the Reference Value (nm)	Standard Uncertainty $u(d_k)$ (mm)	Shortest 95% Coverage Interval	
				Lower Limit (mm)	Upper Limit (mm)
d_1	0.10003	2	0.00020	0.09961	0.10047
d_2	0.20006	−4	0.00038	0.19930	0.20094
d_3	0.30009	−76	0.00037	0.29930	0.30083
d_4	0.40002	−27	0.00053	0.39897	0.40104

Table 4. Measurement results of the glass line scale in horizontal position. Hypothesis 2.

Distances	Estimate d_k (mm)	Deviations to the Reference Value (nm)	Standard Uncertainty $u(d_k)$ (mm)	Shortest 95% Coverage Interval	
				Lower Limit (mm)	Upper Limit (mm)
d_1	0.10002	7	0.00023	0.09960	0.10043
d_2	0.20006	4	0.00043	0.19921	0.20082
d_3	0.30009	−75	0.00044	0.29935	0.30082
d_4	0.40004	−49	0.00058	0.39903	0.40108

If the histograms of Figures 3–5 are compared, it is found that there are great similarities between those obtained in Hypotheses 1 and 3, that is, when histograms are equal to the theoretical distributions obtained from the calibration of the vision subsystem and when they are assimilated to normal distributions. Finally, regardless of the hypothesis considered, if the distance d_k increases, the histogram for this variable approaches a normal distribution.

Table 5. Measurement results of the glass line scale in horizontal position. Hypothesis 3.

Distances	Estimate d_k (mm)	Deviations to the Reference Value (nm)	Standard Uncertainty $u(d_k)$ (mm)	Shortest 95% Coverage Interval	
				Lower Limit (mm)	Upper Limit (mm)
d_1	0.10002	5	0.00020	0.09962	0.10042
d_2	0.20006	0	0.00038	0.19931	0.20081
d_3	0.30009	−75	0.00037	0.29937	0.30080
d_4	0.40002	−34	0.00050	0.39907	0.40102

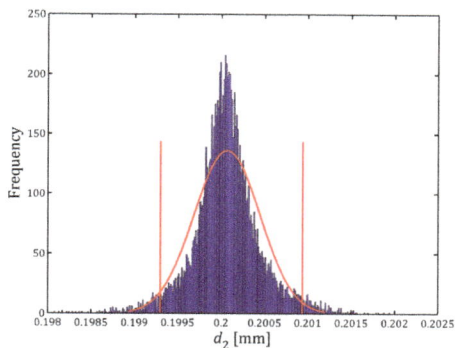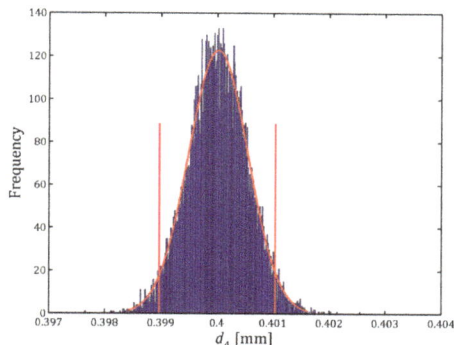

(a)   (b)

Figure 3. Hypothesis 1. (**a**) d_2 histogram; (**b**) d_4 histogram.

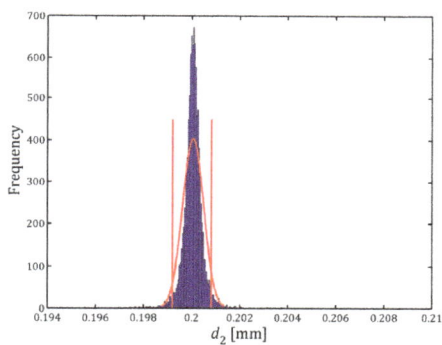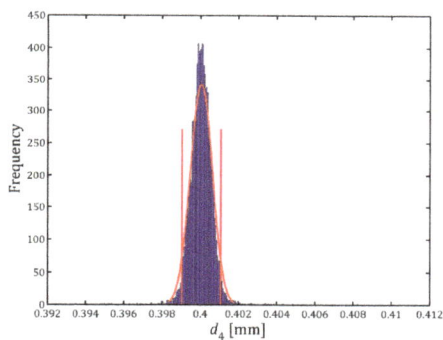

(a)   (b)

Figure 4. Hypothesis 2. (**a**) d_2 histogram; (**b**) d_4 histogram.

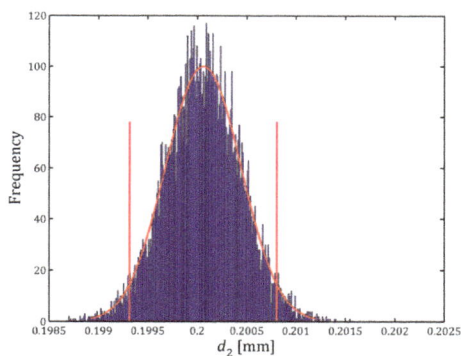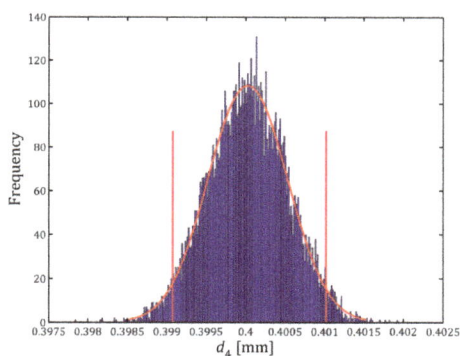

(a)   (b)

Figure 5. Hypothesis 3. (**a**) d_2 histogram; (**b**) d_4 histogram.

4. Conclusions

In this paper, a model for evaluating the influence of the PDFs of input quantities on the output quantities of a measurement model when the measurements are obtained with an optical measuring machine is presented.

The Monte Carlo method is employed to evaluate the uncertainty of the output quantities because this method is the appropriate solution in this type of metrological problem.

When the form of the distributions of input quantities is analysed, in some cases, these do not present known distributions. To employ the Monte Carlo method to calculate the uncertainty in that case, the copula method must be used to generate random correlated data.

Three different hypotheses have been established to analyse the effect of the PDF form in the measurement results. In view of the results obtained, variations lower than 22 nm, it is assumed that the input quantities of the model can be simplified and assimilated to normal distributions.

This allows for a simplifying assumption, reduces the computation time of the programs developed for the subsequent calculation of uncertainties, and reduces the complexity of the code used for generating multivariate random variables, among other benefits. This experimental work may be applied to any model that presents correlated input variables.

Acknowledgments: The authors acknowledge the resources and assistance provided by the staff of the Área de Longitud of Centro Español de Metrología (CEM).

Author Contributions: The work was done in close collaboration. J.C., P.M. and E.G. conducted the experiments, analyzed the data, and wrote the paper.

Conflicts of Interest: The authors declare no conflict of interest.

Abbreviations

The following abbreviations are used in this manuscript:

CMM   coordinate measuring machines
PDF   probability density functions
CEM   Centro Español de Metrología
DOF   degrees of freedom
NPL   National Physical Laboratory
KDE   kernel density estimation

References

1. Kunzmann, H.; Trapet, E.; Wäldele, F. A uniform concept for calibration, acceptance test, and periodic inspection of coordinate measuring machines using reference objects. *CIRP Ann. Manuf. Technol.* **1990**, *39*, 561–564.

2. Schwenke, H.; Siebert, B.; Wäldele, F.; Kunzmann, H. Assessment of uncertainties in dimensional metrology by Monte Carlo simulation: Proposal of a modular and visual software. *CIRP Ann. Manuf. Technol.* **2000**, *49*, 395–398.

3. Wladyslaw, M.S.; Plowucha, W. Analytical estimation of coordinate measurement uncertainty. *Measurement* **2012**, *45*, 2299–2308.

4. Ali, S.H.R.; Buajarern, J. New method and uncertainty estimation for plate dimensions and surface measurements. *J. Phys. Conf. Ser.* **2014**, *483*.

5. BIPM; IEC; IFCC; ILAC; ISO; IUPAC; IUPAP; OIML. Guide to the Expression of Uncertainty in Measurement; JCGM 100:2008 GUM 1995 with Minor Corrections. Available online: http://www.bipm.org/utils/common/documents/jcgm/JCGM_100_2008_E.pdf (accessed on 23 June 2016).

6. Wilhelm, R.; Hocken, R.; Schwenke, H. Task specific uncertainty in coordinate measurement. *CIRP Ann. Manuf. Technol.* **2001**, *50*, 553–563.

7. Abbe, M.; Nara, M.M.; Takamasu, K. Uncertainty evaluation of CMM by modeling with spatial constraint. In Proceedings of the 9th International Symposium on Measurement and Quality Control (ISMQC), Chennai, India, 21–24 November 2007; pp. 121–125.

8. ISO 10360-2: 2009. *Geometrical Product Specifications (GPS)—Acceptance and Reverification Tests for Coordinate Measuring Machines (CMM)—Part 2: CMMs Used for Measuring Linear Dimensions*; International Standards Organization: Geneva, Switzerland, 2009.

9. BIPM; IEC; IFCC; ILAC; ISO; IUPAC; IUPAP; OIML. Supplement 1 to the 'Guide to the Expression of Uncertainty in Measurement'—Propagation of Distributions Using a Monte Carlo method. JCGM 101:2008. Available online: http://www.bipm.org/utils/common/documents/jcgm/JCGM_101_2008_E.pdf (accessed on 23 June 2016).

10. BIPM; IEC; IFCC; ILAC; ISO; IUPAC; IUPAP; OIML. Supplement 2 to the 'Guide to the Expression of Uncertainty in Measurement'—Extension to Any Number of Output Quantities. JCGM 102:2011. Available online: http://www.bipm.org/utils/common/documents/jcgm/JCGM_102_2011_E.pdf (accessed on 23 June 2016).

11. Possolo, A. Copulas for uncertainty analysis. *Metrologia* **2010**, *47*.

12. Papaefthymiou, G.; Kurowicka, G. Using Copulas for Modeling Stochastic Dependence in Power System Uncertainty Analysis. *IEEE Trans. Power Syst.* **2009**, *24*, 40–49.

13. Moazami, S.; Golian, S.; Kavianpour, M.R.; Hong, Y. Uncertainty analysis of bias from satellite rainfall estimates using copula method. *Atmos. Res.* **2014**, *137*, 145–166.

14. Possolo, A.; Elster, C. Evaluating the uncertainty of input quantities in measurement models. *Metrologia* **2014**, *51*.

15. Caja, J.; Gómez, E.; Maresca, P. Optical measuring equipments. Part I: Calibration model and uncertainty estimation. *Precis. Eng. J. Int. Soc. Precis. Eng. Nanotechnol.* **2015**, *40*, 298–304.

16. Caja, J.; Gómez, E.; Maresca, P. Optical measuring equipments. Part II: Measurement traceability and experimental study. *Precis. Eng. J. Int. Soc. Precis. Eng. Nanotechnol.* **2015**, *40*, 305–308.

17. Hartley, R.; Zisserman, A. *Multiple View Geometry in Computer Vision*, 2nd ed.; Cambridge University Press: Cambridge, UK, 2004.

18. Li, D.; Tian, J. An accurate calibration method for a camera with telecentric lenses. *Opt. Lasers Eng.* **2013**, *51*, 538–541.

19. Santo, M.D.; Liguori, C.; Paolillo, A.; Pietrosanto, C. Standard uncertainty evaluation in image-based measurements. *Measurement* **2004**, *36*, 347–358.

20. Santo, M.D.; Liguori, C.; Pietrosanto, A. Uncertainty characterization in image based measurements: A preliminary discussion. *IEEE Trans. Instrum. Meas.* **2000**, *49*, 1101–1107.

21. Nelsen, R.B. *An Introduction to Copulas*, 2nd ed.; Springer-Verlag: New York, NY, USA, 2006.

22. Sheather, S.J.; Jones, M.C. A Reliable Data-Based Bandwidth Selection Method for Kernel Density Estimation. *J. R. Stat. Soc. Ser. B Methodol.* **1991**, *53*, 683–690.

23. Willink, R. Uncertainty analysis by moments for asymmetric variables. *Metrologia* **2006**, *43*.

24. Fotowicz, P. An analytical method for calculating a coverage interval. *Metrologia* **2006**, *43*, 42–45.

MDPI AG

St. Alban-Anlage 66

4052 Basel, Switzerland

Tel. +41 61 683 77 34

Fax +41 61 302 89 18

http://www.mdpi.com

Applied Sciences Editorial Office

E-mail: applsci@mdpi.com

http://www.mdpi.com/journal/applsci

www.ingramcontent.com/pod-product-compliance
Lightning Source LLC
Chambersburg PA
CBHW051921190326
41458CB00026B/6359